I0893256

Teaching Dancing with Ideokinetic Principles

Teaching Dancing with Ideokinetic Principles

Drid Williams

University of Illinois Press
URBANA, CHICAGO, AND SPRINGFIELD

Manufactured in the United States of America
1 2 3 4 5 C P 5 4 3 2 1
∞ This book is printed on acid-free paper.

Library of Congress Cataloging-in-Publication Data
Williams, Drid, 1928–
Teaching dancing with ideokinetic principles / Drid Williams.
p. cm.
ISBN 978-0-252-03608-8 (hardback)
ISBN 978-0-252-07799-9 (paper)
1. Dance—Study and teaching. 2. Dance—Physiological aspects.
3. Human mechanics. 4. Movement, Aesthetics of.
I. Title.
GV1589.W55 2011
792.8071—dc23 2011017748

Contents

Foreword

Dr. Drid Williams provides us with a vivid and perceptive account of the work she did with Dr. Lulu Sweigard, the founder of ideokinesis, during the 1950s and 1960s, and how she applied Sweigard's tenets to her own teaching of dance over many years and venues. This detailed explanation offers an invaluable look at the historical context and groundbreaking teaching concepts that both Williams and Sweigard originated. The field of somatics has expanded dramatically since that time. Both manifestations and inspiration from this seminal work have been developed and applied in different ways, yet it is important to those of us who study and teach in this field to have a clear exegesis of the founder's principles and practices.

A reading of Dr. Sweigard's book *Human Movement Potential: Its Ideokinetic Facilitation* underscores the in-depth and precise level of exposition of which that author was capable. The comprehensiveness of Sweigard's teaching is evident in the clarity of thought transmitted from Sweigard to Williams, as it is outlined in this book.

The subject of ideokinesis first came to my attention in the early 1970s, when I was teaching an introductory dance and movement class for adult beginners in New York City. It was my good fortune to have a great deal of freedom for experimentation in that class. Unfortunately, I often could not find useful ways to assist individual students to improve. Help arrived when I had the chance to take a lesson in ideokinesis with Dr. Williams in 1976, shortly after she returned to the United States after completing her doctoral work at Oxford. That one session set me on a path of study that became central to the rest of my life. My files still hold the piece of paper I saved from that first lesson: 1. Movement begins in the mind; 2. You can't move from a place in your body if you don't know where it is; 3. You will move from where you think that place is, even if you have never consciously thought about it.

Figuring out how to better understand the location, direction, and action of movement within the human body, and then how to provide more accurate images to students, has occupied me for the ensuing thirty years and has become my mind's daily companion. Additional studies in functional anatomy and ideokinesis with Irene Dowd, in breathing coordination with Carl Stough, and in voice with Conrad Osborne have rounded out the ideokinetic approach first offered to me by Dr. Williams. Now that I also teach breathing coordination, I have applied ideokinetic principles to that work. Since none of the deeply internal structures of breathing and voice can be easily observed or touched from the outside, their three-dimensional internal activities respond well to functional imagery.

I did not get to bake biscuits with Dr. Sweigard (as in Chapter 3), but I did get to spend a bitter-cold week with Dr. Williams in Minnesota in the late 1970s, at which time she had sustained an incapacitating foot injury and was simultaneously moving into a new home. I got to pack and unpack, tote, and deliver many cartons up and down stairways while Dr. Williams, with foot appropriately elevated, regaled me with critiques of exactly how and why my body mechanics were not up to carrying out those tasks efficiently. That was just the beginning. It came down to how I did even the most mundane daily chores, including brushing my teeth. The "broadside" approach was not unique to Dr. Sweigard!

Perseverance saw me through. When Dr. Williams came to New York University to establish a master's degree program in the anthropology of human movement (1979–1985), I was able to study ideokinesis in greater detail with her. She supervised my work with students and guided me through practicing and teaching constructive rest. That apprenticeship still informs my teaching of private students and of anatomy for dancers at NYU.

Sweigard's domain deals primarily with the efficiency of the body in its everyday activities. The demands of dance performance often countermand those concepts of efficiency. Sweigard's body of work in ideokinesis cannot be grafted, whole-cloth, onto the teaching of multi-varying dance techniques. Indeed, Sweigard herself counseled against such an attempt. Williams has understood the difference wisely and well, as the reader will soon see.

Lynn Martin
New York City

Introduction

When I studied with Dr. Lulu E. Sweigard (b. 1895–d. 1974) in the late 1950s, I went to her for "neuromuscular reeducation." The name "ideokinesis" didn't yet exist (the word is a combined form meaning *ideo* = "idea," plus *kinesis* = "body movements" or "gestures"). This name was applied to the discipline in the early nineteen-seventies. There were comparatively few people in the New York dance world (or elsewhere) fifty years ago who had heard of Sweigard and her work in movement education. Moreover, "neuromuscular reeducation" was not an attractive name, because it implied that something was wrong—something had to be "*re*educated." Nevertheless, I sought her out, and thereafter, my teaching of dancing was changed in fundamental ways.

According to Pamela Matt (professor emerita, Arizona State University), I am one of the second generation of teachers inspired by Sweigard's work, but I would not have gone to her at all if I hadn't been deeply interested in teaching dancing without physical injury to students. This aim dominated my approach to dance teaching from 1948 until my retirement from teaching dancing in 1970. In fact, the chapters of this book were originally letters addressed to Pamela Matt, who, when our correspondence began in August 2003, was still teaching dance science and somatics in the dance department at Arizona State. We began our extended dialogue as Professor Matt was developing a website, www.ideokinesis.com, for ideokinesis, now available for study.

Because I was Sweigard's student in the late fifties, Pam wanted to know how I used her work both during and after the time I studied with her. In particular, she was interested in connections with the teaching of dancing, for as she points out:

> Most of the students of Clark and Sweigard were dancers or were affiliated in some way with the dance profession. Some of them continued to teach the material largely as they had been taught, but many others applied knowledge they attained in their study of ideokinesis to the teaching of [the] dance. Thus, the second era in the development of ideokinesis was shaped by the *Dance Generation*. (From the "Introduction," page 2, of www.ideokinesis.com)

In this book, I attempt to make as clear as possible what I learned from Sweigard. I also aim to make clear the differences between teaching Sweigardian ideokinesis and teaching any form of dancing by pointing out in the first chapter how I used Sweigard's *principles* in teaching dancing but did not use her images, nor did I teach "ideokinesis" in ballet or modern dance classes, which permits me to say that the synopsis of chapters and appendices of this book summarizes its content.

However, before looking at the synopsis of chapters, I want readers to know what Martha Hill (director of the Dance Department at Juilliard School of Music for many years) said in an introduction to an article written by Sweigard for *Impulse Magazine* (reprinted as the Appendix to Chapter 3, page 47):

> To anyone who is familiar with the historical development of research and practice in the field of posture and efficiency in movement, Dr. Sweigard's work has always been in the vanguard of the mid-twentieth century revolution in this field. Soundly based in the sciences of anatomy and body mechanics as well as neurology and physiology and drawing upon developments in psychology, her principles provide a common grammar for dancers and teachers of dance:
>
> "Movement resides in the thinking, not muscle action."
> "There is no voluntary control over muscle action."
> "Thinking and emotions influence muscle action."
> "Movement imagined in the body results in change in established patterns of neuro-muscular co-ordination." (Hill 1961: 38)

Some—or all—of the statements Hill lists are as hard for some people to believe today as they were in 1957 when I met Dr. Sweigard, especially the second: "There is no voluntary control over muscle action," a statement that is true if (and only if) it is understood in terms of Sweigard's emphasis on *posture,* that is, "the alignment of the body as it occurs habitually in the standing position with the weight distributed as evenly as possible on both feet" (Sweigard 1961: 39), and if it is understood that subcortical controls in the nervous system determine the habitual postural patterns of all human beings.

Synopsis of Chapters and Appendices

Chapter 1 ("Beams of Light") discusses how I used (and didn't use) Sweigard's imagery in relation to teaching dance classes. The necessity of working from the *principles* involved in Sweigard's imagery in classes, in contrast to *direct applications* of her imagery to a dancing class, is accentuated. "Imagery" is separated into two types: 1. "anatomical," relating directly to the body, and

2. "aesthetic," pertaining to the execution of danced moves. Throughout, I emphasize teaching dancers to protect themselves from bodily injury in dance classes and rehearsals.

The Appendix to Chapter 1 ("Lexicons of the Body") is a brief clarification of taxonomies of the body and their potential influence on dance teaching.

Chapter 2 ("Relaxation") is about Sweigard's strong distinction between ordinary relaxation and *tension reduction* in the constructive rest position (CRP). Some aspects of the philosophical dissimilarities between Sweigard's ideokinesis and the kinesthetic legacy of Barbara Clark are discussed. By way of comparison, I introduce Bob Fleshman's translation of Sweigard's imagery into mime classes and the problems generated from such translations.

The Appendix to Chapter 2 ("Constructive Rest") consists of an early article where Dr. Sweigard explains how using images in the constructive rest position assists change in postural habits, reducing strain and developing concentration through the use of mental imagery. A useful exercise is given to accomplish these purposes.

Chapter 3 ("Baking Biscuits and Kinesthesia") speaks to the differences between teaching ideokinetic techniques and awareness to *non*dancers and to dancers; the definition of *kinesthesia;* and a direct versus an indirect teaching of kinesthetic awareness and related concepts. There is a brief discussion of two books, *This Magic Body* (Lee 1946) and *The Hidden You* (Todd 1953).

The Appendix to Chapter 3 ("The Dancer's Posture") is an article by Sweigard that pertains to flexibility tests used in her posture laboratory at Juilliard and her analysis of the relationship of posture to dancing.

Chapter 4 ("Doctors, Dancing, and Ideokinesis") deals with the concept of normalcy regarding movement, and the medical profession's warrant to cure the body, *versus* the teaching of dancing, which typically begins where medics leave off. In this chapter, I describe three professional dancers' problems and their rehabilitation, and the chapter ends with the limitations of classroom techniques in ballet dancing.

The Appendix to Chapter 4 ("Better Dancing through Better Body Balance") was written specifically for dancers, concentrating on "turn out" of the legs and feet; this article by Dr. Sweigard seeks to explain some of the difficulties that a dance technique can produce in a dancer's body. The sections on analysis and the remedies she suggests are particularly helpful.

Chapter 5 ("Mirror, Mirror . . .") discusses mirror images and their deleterious effects upon dance teachers and students of dancing. This chapter also makes a transition between the easily understood concept of "mirror image" and less understood concepts of image and their use with regard to movement—in effect, "Mental Practice."

The Appendix to Chapter 5 ("Accentuate the Positive . . .") deals with a major problem connected with teachers' verbal delivery of images in the classroom.

In Chapter 6 ("Imagery and Habit"), I attempt to describe four different kinds of imagery associated with the teaching of dancing, in particular emphasizing the differences between *internal* (kinesthetic) imagery and *external* (mirror) imagery.

The Appendix to Chapter 6 ("Walking Bundles of Habit") is from volume 1 of William James's *Psychology,* first published in 1890. This excerpt from James's writing is meant to emphasize the fact that a dance technique class is a "development of habits" class. Few dancers recognize the fact (and the force) of habit development in their training, although they would be wise to do so.

Chapter 7 ("More about Teaching Dancing") examines the problem of body types that appear in all classes—mainly that of differing bodily types (endomorphs, mesomorphs, and ectomorphs) in the same class. Jean-Georges Noverre's exposition of *arqué* and *jarreté* dancers and their needs are introduced and explained.

The Appendix to Chapter 7 is Letter XI from "Letters on Dancing and the Ballet" by Jean-Georges Noverre (1930 [1760]: 109–16). Written roughly two and a half centuries ago, this letter from Noverre is outstanding for his discussion of *arqué* and *jarreté* legs—characteristic of two major bodily types (and their variations) to be found in any dance class. Students and teachers alike will profit from the knowledge and advice given so long ago.

Two Final Notes

The restricted scope of this book (teaching dancing, imagery, and ideokinesis) and the language in which it is written prevents examination of critical issues accompanying the *general usage* of imagery—for example, the many clues contained in Brenda Farnell's excellent essay *Metaphors We Move By* (1996), a definitive post-Cartesian rejoinder to George Lakoff and Mark Johnson's book *Metaphors We Live By* (1980), in which the authors construct a notion of a "kinesthetic image schema" that aims to establish a basic level of physical experience that is preconceptual (see Farnell 1996: 311).

In a practical book such as this, one is required to leave out vital intellectual and theoretical issues that usually tend not to arise in day-to-day teaching. For example, the distinction I make between anatomical and aesthetic imagery in Chapter 6 is admittedly arbitrary. As anthropological colleagues will recognize, there are elements of metaphor in both categories; however, the distinction (and why I used it in dance classes) will become clearer when the "passive/active" opposition is considered. "Aesthetic imagery," as I used it, was *always* utilized when *movement* was involved in sequences of danced movements.

The Appendix to Chapter 1 ("Lexicons of the Body") points to important anthropological issues that would be helpful to dancers and teachers of dancing, if they were understood, and there are others: 1. the notion of universality (Williams 2004, with particular reference to the essay "An Exercise in Applied Personal Anthropology," in Williams 2004: 231–54); 2. the notions of "primitive dance," indigenous classifications, and ethnocentrism (Farnell 1999: 145–60); 3. the notion of "thinking with movement" (Hart-Johnson and Puri 1995: 158–86); and 4. "The Indexical Structure of Visibility" (Urciuoli 1995: 189–215).

Finally, I want to thank Dixie Durr for the meticulous pre-reading that she did of the manuscript. Most of all, I would like to acknowledge the patience and inspiration lavishly provided me over the past three years by Pamela Matt, whose comments, questions, and criticisms have been invaluable throughout. This book is dedicated to her and to her lifelong struggle to promote understanding among highly diverse (often contentious) groups of "movers," not all of whom care to cooperate with (or even listen to) one another.

Chapter 1
Beams of Light

Because I took courses in physiology and anatomy at Jefferson High School in Portland, Oregon, during the early forties, Dr. Sweigard said I could begin lessons with her after completing a single private course in bone anatomy with Dr. C. A. de Vere in New York City. I enrolled with him in 1957 and worked with him for six months. I started with Dr. Sweigard at the end of 1957 and continued with her for approximately a year and a half to two years.

Attending the lessons was difficult, because at the time, she was not teaching at Juilliard full-time; therefore she did not often come to New York. Most of the time, I traveled to her home north of New York City in Tomkins Cove. I had lessons with her more or less once a week for the first six months, and after that, roughly twice a month (every other week) for a year.

She visited my studio to watch me teach a dance class during the first month I was with her. That visit produced a lecture on the imagery I used in dance classes best described by saying that she went through the teaching imagery I used "like Sherman through Georgia." She was scathingly critical of nearly everything I said in dance classes, justifying her criticisms of my verbal performance by saying that many of the images I used were either anatomically incorrect or that they focused on the aesthetics of the dance form I was teaching, not on the body itself. She also said that the imagery I used was "unexamined—that is, I hadn't really thought it out, despite my anatomical training. She also said that the imagery I used in classes simply repeated what I heard in classes when I learned to dance.

It was from her criticisms of anatomically incorrect imagery that I first got the idea of separating two categories of images pertaining to dance classes, such as "anatomical" and "aesthetic." She was not prepared to tell me anything about what I shall hereafter call "dance imagery" (the aesthetic kind), because, she said, she knew nothing about dancing and was, frankly, not interested in it, except for working with dancers on their postural and mechanical problems.

Any anatomically based imagery I did use that alluded to bodily parts, such as "Keep your spine straight," "Point your toes," and the like, she totally destroyed by pointing out, for example, that what I meant by "Keep your spine straight" was "Keep your axis of gravity straight." She also pointed out that the toes themselves did *not* "point." The correct image pertains to the ankle joint in order to achieve plantar flexion of the foot. The ankle joint should be extended to the maximum possible, the toes themselves stretching out on a line and kept straight. There was to be no "clenching" of toes—a habit common among dancers and gymnasts that frequently requires extended retraining.

She was convinced that one of the basic problems for most dancers consisted of the fact that they had no idea where movements really come from in the body. They think in terms of generalities such as "hip," "thigh," "shoulder," and "back." They think of *the outer contours* of these bodily parts. Years later, after studying social anthropology, I discovered that dance teachers (in common with everyone) use a "social lexicon" of the body (see Appendix to this chapter, pages 18–20) that, for the teaching of movement, has substantial negative consequences.

Especially beginning dancers tend to think of the external shapes of their bodily parts such as "hip" instead of thinking in terms of the bony structures and connections in the body, say, "thigh joint," for instance, which locates the origin of most of the movements of the legs, or the "gleno-humeral joint," which locates the origin of nearly all arm movements. Rote learning of the anatomical terminology doesn't help. Learning anatomical terms should mean learning elements of functional anatomy that are important for the teaching of dancing.

To Sweigard, the basic problem with most dance teachers was that their use of imagery in classes was "received." They tend simply to repeat whatever they learned from whoever taught the classes they took, thus the usage of incorrect or distorted bodily imagery is transmitted from teacher to student, and has been for centuries. I do not exaggerate. According to Derek Lynham, Jean George Noverre was aware of such problems in the mid-seventeen-hundreds:

> Granted that the would-be dancer is suited by nature to his profession, the first duty of the teacher is to study [the student's] physical conformation and his [or her] temperament and to instruct him [or her] accordingly. "There may be only one right principle to be taught, but is there only one way of demonstrating it and of imparting it to the students one undertakes to teach, and must one not of necessity lead them to the same end by different ways? I admit that in order to do so *one must have real knowledge, for without thought and study it is impossible to apply given principles varied to suit different types of physical conformation and varying degrees of aptitude,* for one cannot see at a glance what

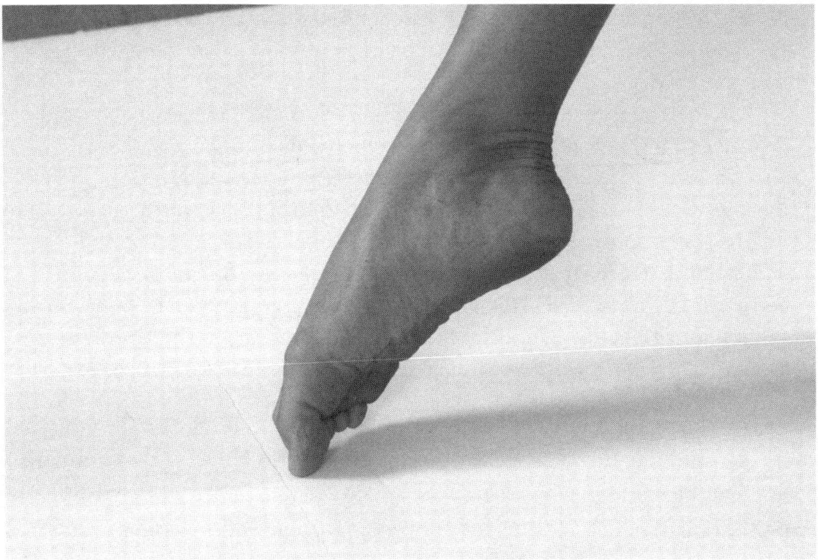

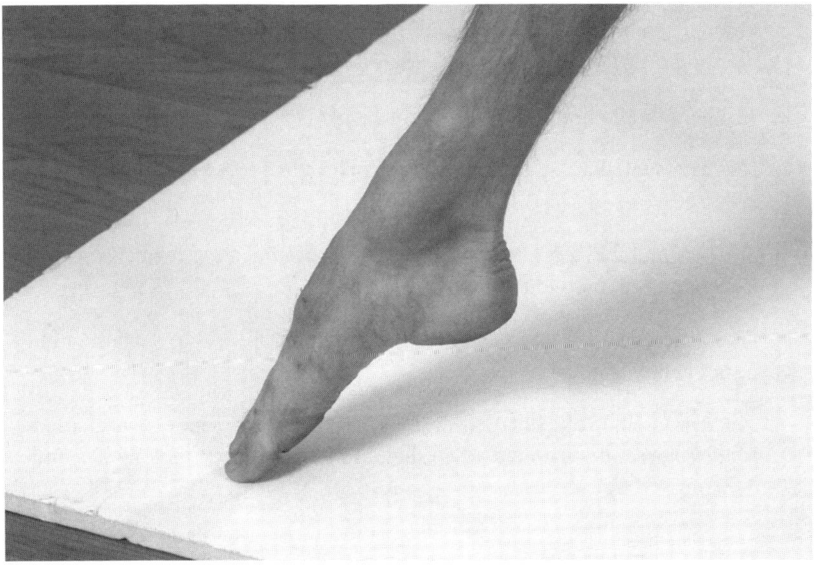

Images 1 and 2. The correct image pertains to the ankle joint in order to achieve plantar flexion of the foot. The ankle joint should be extended to the maximum possible, the toes themselves stretching out on a line and kept straight (2). There was to be no "clenching" of toes (1)—a habit common among dancers and gymnasts that frequently requires extended retraining. © Human Kinetics

> is convenient to one and cannot suit another, and one cannot vary one's classes in proportion to the variations which nature and habit, often more rebellious than nature, offer and present us.
>
> "It is, therefore, left to the teacher to train each student in the style to which he is suited. It is not sufficient to possess a thorough knowledge of the art; it is also necessary to guard against vain pride which inclines each one to believe that his [or her] manner of execution is the one and only one which can please, for a teacher who always puts himself forward as a model of perfection and who endeavours but to make of his students a copy of which he himself is the good or bad original, will succeed in making tolerable dancers only when he meets with subjects gifted with the same disposition, height, conformation and intelligence as himself." (Noverre 1930[1760], cited in Lynham 1950: 138—italics added)

I ask readers' indulgence for the lengthy quotation: it may be the case that there are students nowadays who do not know who Noverre was. I am also aware that the world in which I grew up and learned to dance was very different from the one in which we now live; however, this quotation from a renowned teacher over two hundred and fifty years ago makes six important points:

1. Teachers are told to study students' physical conformations, that is, *to understand their postural characteristics.*
2. Teachers should find *many different ways* of teaching a right principle.
3. Noverre says that following his program of teaching requires *real knowledge*, and
4. Teachers must be clever, because they don't teach students individually—they teach *groups* of individuals.
5. Finally, teachers are responsible for teaching each student to the best of his or her ability, and
6. Teachers shouldn't be full of pride, believing that *their* manner of execution is perfect, secretly flattered when their students turn out looking like and dancing as they do.

Applications of Sweigard's Work

On the whole, my impression was (and still is) that Dr. Sweigard didn't think much of the plight of dancers in their quest for learning and/or excelling at some danced idiom, although I would not want this remark to be misconstrued. She was against *any* system of physical training, including gymnastics, military drill, or any form of dancing that proceeded from a "held" posture, a hyperextended chest or the "holding" of any set of muscles. Then, too, she didn't think much of how ballet and modern concert dancing were taught, even though she was allied with the Juilliard school. Insofar as I experienced her teaching, *her entire focus was on the habitual posture of individuals in the constructive rest position, in a sitting position, in an ordinary standing position, and walking*. These "postures" were of paramount importance to her:

> . . . posture in the standing position is a dynamic phenomenon in which the amount and extent of muscle work, and the wear and tear on the skeletal framework and its joints and ligaments, depend largely on the efficiency of the neuromuscular coordination engaged habitually in maintaining upright balance. (Sweigard 1974: 174)

The *posture* with which an aspiring dancer started was the all-important factor. Many dance students and teachers found (and find) this principle unacceptable.

I had no argument with Sweigard's thesis, for I could see that improvement of the postural pattern of each individual student was the only way *permanently* to improve his or her dancing. However, I wanted to know what one taught with regard to teaching idioms of the dance, the nature of which often requires holding the body in various positions as part of the aesthetics of the dance form. Sweigard refused to discuss that kind of thing with me, thus I had to work out how to handle the problem on my own. I thought it was a pity that she didn't work more with dance teachers, for it is teachers who need to know that

> To illustrate the ideokinetic power of imagination on subcortical planning of muscle action, *visualize yourself growing tall* and, in so doing, stand upright with greater ease. If you *forcefully stretch up* to make yourself taller, however, you will find your body taut and unyielding. In the former procedure, your muscles responded to subcortical directions, without interference of conditioned neuromuscular reflexes which may be inefficient. In the latter case, *your consciously directed movement used established neuromuscular habit patterns* which eliminated to a marked degree the possible influence of subcortical planning. (Sweigard 1974: 170—italics added)

Knowing this kind of thing—really understanding it—means that teachers will avoid using directions such as "Pull up," "Stretch up," and the like while teaching dance technique classes.

In my case, however, I didn't try to apply anything I learned from Sweigard during the first year I studied with her, *apart from changes in the general bodily imagery I used in classes*, such as "Keep your spine straight," or "Turn out your knees"—both of which are wrong—and any other images that used *direct* anatomical or bodily referents.

She was adamant that most neuromuscular reeducation had to take place on an individual level, and she warned that she didn't want to find out that I was teaching anything I learned from her on an individual basis. She threatened to terminate my classes with her if she got the idea that I was cheating. Over and over, she stressed the fact that my pattern of bodily problems was mine, *and mine alone.* As far as I know, the same held true for every individual with whom she worked. Even after I had passed the test, so to speak, I didn't incorporate any of what I learned from her as an individual into my classes. I will say that, apart from the generalized imagery regarding bodily parts described above, I devised a series of "techniques" (I can find no better word), usually given at the beginning of class, to get dancers to think about their bodies more effectively.

Procedures

For example, I would have the class sit on the floor, legs straight out in front of them, and then I would say, "Now, *without moving your legs at all,* turn out your feet." Of course this is impossible. Then I would say, "Turn out your knees." They couldn't do that, either. Next I would say, "Turn out your legs from your thigh joints." Of course they could do that, following which I would explain *why* this was anatomically true, hammering home the point that "turn-out" occurs *only* at the thigh joints. All the knees and feet are meant to do is to follow. As we all know, however, many dance teachers use the images of "turn out your feet" and "turn out your knees" in classes, but both images create misconceptions of how the body works and they reflect anatomical impossibilities.

In other words, there are ways in which the body moves that aspiring dancers (even professionals) simply don't know about, starting with the fact that knees and ankles have comparatively limited movements on their own. Most danced movements in any idiom of dancing come from the thigh joints. Similarly, elbows, wrists, and finger joints have comparatively limited movements (except in eastern dance forms such as Balinese and Javanese dancing and south Indian Bharatanatyam) in contrast to the gleno-humeral complex.

Mind you, *none* of this (effective as it was and still is, by those who use it), strictly speaking, is Sweigard's imagery, or her use of images to change the neuromuscular habits of individual bodies. I cannot stress that point too much.

The body techniques I taught in dance classes were my own, and I had such techniques for every part of the body. I would only give one—perhaps two—at the beginning of each class because I discovered that most students couldn't really absorb more than one or two at a time, and they wanted to get on with the dancing. As Sweigard says, "Traditional concepts and loyalties die hard" (1974: 5).

Using such procedures as I did, I could then refer to the students' thigh joints (or some other joint) during the dance class and students had some means of understanding what I meant for them to do. Realize, however, that these techniques were mainly about the *locations* of key skeletal parts of their anatomy (mainly the location of joints). Their ability to locate these points in their own bodies is crucial to a dancer's success. I explained that *they had to know where important anatomical points were located in their own bodies.* They learned to locate these (with their own fingers making pressure points) during work at the *barre,* or in floor work.

It is interesting, I think, to ask a beginning (or even advanced) class of modern or ballet dancers to locate their thigh joints for you at the beginning of a class. They will usually put their thumbs or index fingers on the iliac crests, or over the great trochanters of the femurs. Sometimes one of them might put his or her thumbs on the front of their bodies over the thigh *joints,* but that is rare. From the mistakes (because everyone looks around to see what everyone in the class is doing), you are then, as teacher, in a position to explain where the thigh joints really are, and you're off and running (or dancing!). I also told students what *wrong* imagery amounted to, and at those times, I did repeat Sweigard's words, telling them, for example, that they *could not* "straighten their spines," "point their toes," or "turn their knees out," or, if they *did* accomplish what these images suggest, they would be in serious trouble.

Gradually, the dancers I trained came to realize that they were personally responsible for their own work in any kind of dance class they took. They soon began to recognize that many teachers of dancing know little or nothing about what they are doing from this more anatomical point of view. However, they found they could compensate for lacks in other teachers' knowledge by substituting anatomically correct images for those they were given in classes.

For example, if a teacher told them to "point their toes," they now knew what they *really* had to do: extend their ankles to the maximum and stretch their feet (including the toes) away from the ankle. In other words, *they learned to protect themselves* from poor bodily imagery in dance classes (see Sparger 1982[1965]: 44–45).

I'm happy to say that I never injured a dancer in any of my classes, and, because those I taught knew how to protect themselves, they didn't injure themselves in other teachers' classes, either. I've only given a couple of examples of poor imagery here, of course, but there are many: "Tuck your butt [or "rear" or *derriere*] under" is a common example, because, of course, with

Images 3 and 4. It is interesting, I think, to ask a beginning (or even advanced) class of modern or ballet dancers to locate their thigh joints for you at the beginning of a class. They will usually put their thumbs or index fingers on the iliac crests, or over the great trochanters of the femurs (3). Sometimes one of them might put his or her thumbs on the front of the body over the thigh joints (4), but that is rare.

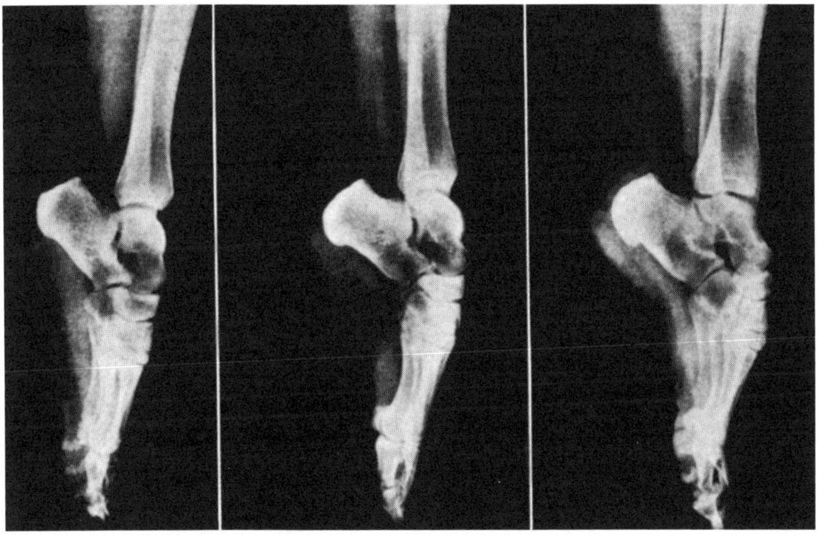

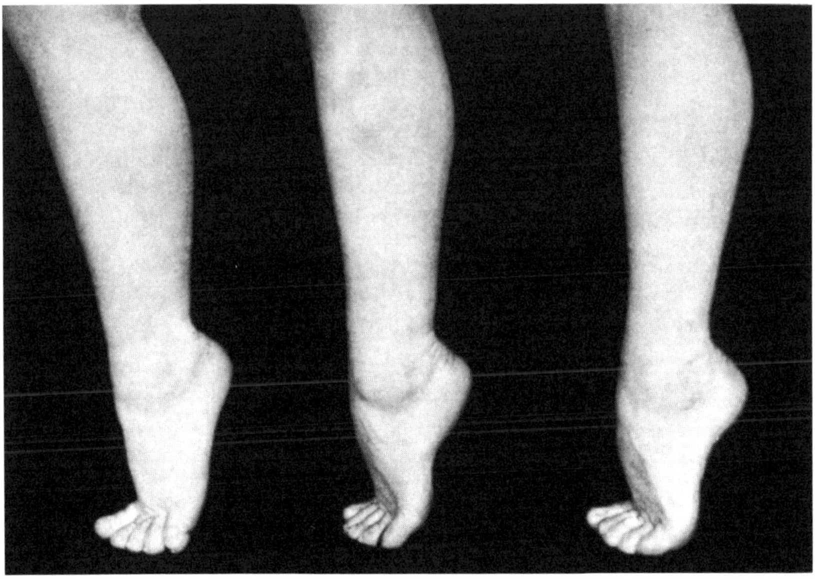

Images 5 and 6. Three types of feet, bare (6) and X-rayed in *pointe* shoes (5): 1. Strongly knit with little arch; 2. High arch from ankle joint with long forefoot; 3. High arch with more prominent tarsal bones (from Sparger, pp. 44–45).

the buttocks clenched, no movement (even ordinary walking) is possible.* I went through every image I used in dance classes that had bodily imagery connected with it (not repeated here), and I suggest that teachers engage in this kind of exercise for themselves. If the exercise is conscientiously carried out, it changes the whole verbal structure of the class, ultimately changing how students perform.

Teaching Dancers to Handle Their Own Bodies

Because everyone knows that dance teaching (as everything else) is controversial, it must be said that there were a few students who objected to my post-Sweigardian methods of teaching. They said, "This isn't *dancing*, and that's what I came to you to learn." Students who didn't want to learn about their bodies were in the minority, but from their criticisms we can, I think, tease out some important pedagogical points.

First, "technique classes" in dancing should provide for student dancers learning something about their own bodies—not in the usual fashion of taking anatomy, kinesiology, and dissection classes. With that kind of detailed knowledge they are often lost because they have no idea how to apply the knowledge to their own or anyone else's body, far less to an idiom of dancing. Eventually, because of the study of ideokinesis, I went through the whole repertoire of my own teaching practices and, as a result, worked out what it was that a ballet dancer or modern concert dancer needed to know about anatomy to function safely and more effectively in dance classes.

Specifically, dance students don't need to imagine a skeleton or bones when they are dancing, but they *do* need to know the key points in their own bodies from which movement originates. When the use of these key points becomes habitual, they don't "think" about them at all. They are able simply to perform what is required. Teaching this kind of thing should happen during a dance class, preferably at the beginning, following which regular training in dance techniques or forms can take place.

Lest anyone think this is a new idea in the teaching of ballet, I will point again to Noverre, who said, "In my youth, I studied osteology which I have found of great advantage in my teaching, both to allay the tediousness of the lessons and to throw more light on the demonstration of principles. This art has taught me to see the causes which oppose themselves to the execution of this or that movement and, knowing the bony structure of [people] and

*Some teachers use the image of putting a quarter (coin) between the cheeks of the buttocks—a particularly damaging idea, apart from the fact that you cannot walk!

the levers and hinges which govern its play, I did not require from my pupils that which nature did not allow and I directed their classes after a deep analysis of the conformation of each one of them" (Noverre 1930[1760], cited in Lynham 1950: 138). The idea of teaching movement through the use of imagery is not unknown in other professions, either. The concept was used in teaching piano technique in the early fifties (see Bonpensierre 1953) and violin technique (Havas 1964).

Second, I think dance teachers need to have some of the traditional knowledge that bone anatomists, kinesiologists, and physiologists possess, but I would limit such knowledge to the specifics that pertain to a living, *moving* body. I took anatomy and dissection classes in 1961 at the University of Wisconsin, Madison, where I did undergraduate study in philosophy. An anecdote about a conversation I had with the professor is relevant: we had just completed a midterm examination. She called me to her office to discuss the exam, saying that she was perplexed about what grade to give me because half of the exam she considered "brilliant" and A-level work, but the other half was nowhere near that standard. She wanted to know why. I told her that the part that was good was because the knowledge pertained to *a moving body* and the teaching thereof. The rest of the exam was primarily memorization and irrelevant to the teaching of dancing. I am told that making judgments of this kind are not necessary today because the same attitudes don't pertain; however, she said, "We can't teach anatomy classes that are so narrowly focused." I said I was aware of the problem, and that knowledge, for example, pertaining to the condyles of the leg bones, joint surfaces, and so forth (part of the exam I did not do well) was undoubtedly important to physiotherapists, surgeons, and others in those lines of work, but for dance teachers, such knowledge was irrelevant. She gave me a B-minus on the paper, but the real problem was that *the dance department never made any effort to design or teach a class in anatomy that was useful to teachers of dancing.* I am told that this is no longer the case: there are at present functional anatomy classes in good dance departments that are geared toward helping dancers accurately understand the real shapes of bony joints in the body, their potential and limitations.

That point leads to a third: anatomical knowledge *in the abstract* isn't useful unless the learners know how to apply what they may learn to their chosen profession or line of work. If a physiotherapist has to teach a worker to compensate for, say, an accident in a steel mill that has removed half his foot, then the therapist needs to know in minute detail the condyles, tendonous attachments, ligamentous attachments, and so on of the ankle or foot, because she or he has to get the worker's limb somehow to compensate for the disability in ordinary, everyday movement, but dance teachers rarely teach people with such disabilities, do they? Someone who has suffered severe trauma to a foot, knee, thigh joint, or spine isn't going to be able to dance, so why attempt to turn dance teachers into physiotherapists? If this seems a bit strong, I want to say that teaching such abstract knowledge as anatomy represents should be guided by the purposes it will serve. These days, there are centers where physiotherapists have been taught to recognize and deal with injuries common to the dance, such as the Harkness Center for Dance Injuries, at the Hospital for Joint Diseases in Manhattan, but dancing itself is not taught there.

Fourth, there are many people who take dancing lessons who will never achieve professional-level performance abilities. This does not mean that they shouldn't develop their movement potential to the highest possible level, but we return to this subject in Chapter 4. Now we will get back to the use of Sweigardian *principles* regarding imagery in dance technique classes.

I always explained to dissenters in my classes that they couldn't dance very well if they didn't know *basic locational facts* about their bodies (example: knowing where their thigh joints are). This is because the human nervous system is such that messages from the brain go to *sets* of muscles in the body. If one thinks or "images" the thigh joint, say, at the iliac crest (instead of the ilio-femoral joint—(see Images 3 and 4), then the required leg movement is bound to be wrong, because the iliac crest *is not* where the thigh joint is located. The messages created by anatomically incorrect images are sent to the wrong place in the student's body, and distorted, poor, or failed moves result. It's like dialing a wrong telephone number—wrong number, wrong person. Wrong images produce faulty movement responses. It really is as simple as that.

I would, however, like to add this comment: for those students who refused, even after reasonable explanation, to accept the kind of teaching I did, I simply suggested going somewhere else to study, because they hindered those in classes who wanted to learn what I taught. Not everyone responds positively to the kind of teaching I advocate, but the students I gained far outnumbered the ones I lost, so I never regretted the decision to let dissenting students go.

Working from Sweigard's Principles, Not Her Imagery

Apart from locational techniques, another major use I made of the principles involved in Sweigard's work in dance classes was the use of light beam imagery, including the axis of gravity. To my knowledge, Sweigard herself never used light beams as an image. She had various images for the lines of movement, or lines of action in the body, but, as far as I know, light beams were not among them (see Sweigard 1974: 254–57, 'Standing' and 'Walking').

After students had learned key locational points in their bodies, I would (in the portion of the class I devoted to such procedures) get them to work in pairs to establish straight lines (in the form of pencil-thin light beams)* from the head of the humerus, to the center of the elbow, to the center of the wrist, and out the second finger of each hand.† These were *not* specifically Sweigard's "lines of movement" or "lines of direction" such as she used in individual sessions, nor are those interchangeable with the light beams. (See Sweigard 1974: 237–45, for some of her imagery of these.)

The light beams were meant to be centering images *based upon* (but not the same as) the lines of movement ("force" or "direction") in the body. They were useful in teaching the aesthetics of a dance form, as well, a lot of which (especially in ballet) consists of geometric design. The main purpose of the light beams was to get whatever movement the student did *centered in the joints and surrounding musculature* of his or her body.

To teach the light beams in classes, I used the buddy system (working in pairs). With thumb and index finger, one of the students would help his or her partner locate the center of the gleno-humeral joint (the shoulder) by applying a light pressure on the front and back of the joint. Then the partner would "find the center between the two pressure points." Next, the first student would locate pressure points on the medial and lateral points of the elbow joint and the partner would imagine the center point there. Now the first student would apply both sets of pressure points at the same time (between head of humerus and elbow). The partner would picture (imagine) a pencil-thin light beam going from the head of the humerus to the elbow through the center of the upper arm. The lines from the heads of the humerus can be connected through the upper chest, thereby teaching as if the arms come from the spine.

I then gave some simple arm movements—"raise the upper arm to the side to shoulder height," or "move the upper arm to the front" (again, shoulder height), or whatever—while imagining the light beam (*not* the "arm") moving in space. The student who had applied the pressure points simply watched.

---

*"Laser beams" would work just as well.

†I use the arm as an example because it is the easiest to deal with in writing. These procedures using the light beams were carried out in my classes for legs, torso (axis of gravity), and head. NB: The "hand" is here thought about as having one thumb and four fingers, thus the "second finger" is next to the index finger. To some, the "hand" has five fingers (including the thumb), thus the light beam, in this conception, passes through the third finger.

Image 7. To teach the light beams in classes, I used the "buddy system" (working in pairs). With thumb and index finger, one of the students would help his or her partner locate the center of the gleno-humeral joint (the shoulder) by applying a light pressure on the front and back of the joint. Then the partner would "find the center between the two pressure points." Next, the first student would locate pressure points on the medial and lateral points of the elbow joint and the partner would imagine the center point there. Now the first student would apply both sets of pressure points at the same time (between head of humerus and elbow). © Human Kinetics

It is immediately apparent to someone teaching this kind of exercise that the *quality* of the movement, even of a single bodily part, is different—and much better. It is also obvious to students that the movement is improved. It is, in their descriptions, more centered. It is easier. What the students are doing, of course, is taking their minds off whatever they think their arms look like, or what they look like to themselves in a mirror, or some such nonsense, because *the image centers the muscle action around the appropriate joint(s)*. We would repeat this process several times, so that I was sure that the "centers" had been well and truly established. We would then situate a longer line through the arm by adding the center of the wrist, then finally the hand.

With the light beam only in the upper arm and forearm, the students made simple angles in space. We next progressed to curves (as the arm is

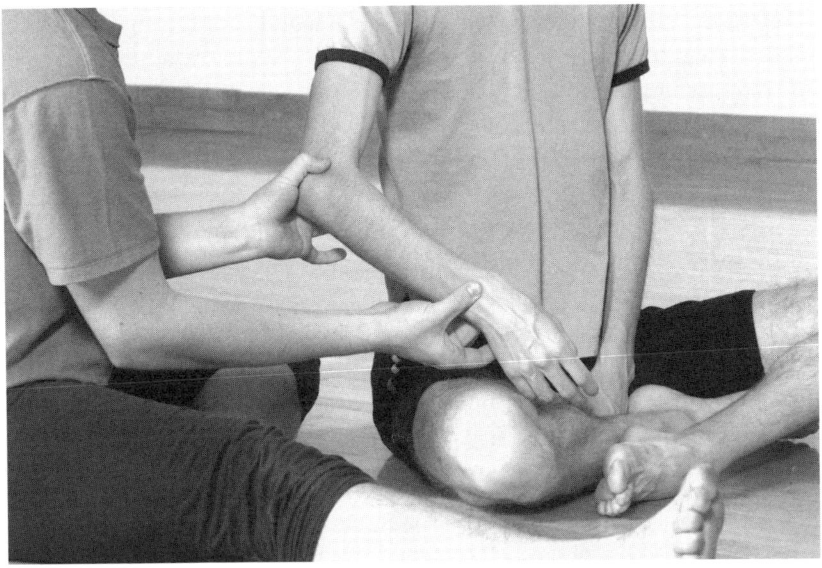

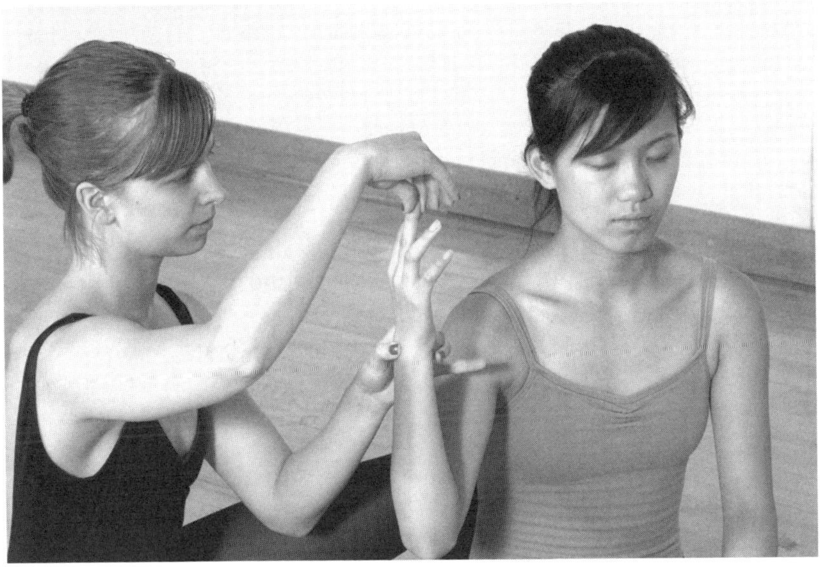

Images 8 and 9. We would repeat this process several times, so that I was sure that the "centers" had been well and truly established. We would then situate a longer line through the arm by adding the center of the wrist, then finally the hand. With the light beam only in the upper arm and forearm, the students made simple angles in space. We next progressed to curves (as the arm is in ballet dancing). Finally, when we added a light beam through the hand . . . © Human Kinetics in ballet dancing). Finally, when we added a light beam through the hand, students had the whole limb to work with, and I assure you, the change in the movement was noticeably improved. Why?

Because the light beams center the muscular action in relation to the joint(s), thus tending to function with maximum efficiency. Again, it's that simple. But it is also complex, because the student must discipline himself or herself to *think of the light beam moving, not* the arm.

Throughout my own study of anatomy and the work with Sweigard, I learned that the commonplace notion of "one muscle–one movement" is wrong. Even a simple movement, such as flexing an arm, involves the action of *many* muscles. The muscular actions that achieve danced movements are too complicated to be of assistance in teaching dancing (or, in my opinion, any movement form). Celia Sparger says it very well:

> Anatomy is a very large subject and it is a science. Dancing is an art. The scientist has knowledge of the living body; the artist of the moving body. The dancer's knowledge of the moving body is often more detailed and profound than anything to be found in science, but she has not the language in which to express it. She knows by feeling, and by experience within her own body what the scientist knows by study from the outside. It would be a loss should she try to substitute the one for the other. . . . Curiously enough, anatomy seems to suggest to most people the study of muscles. Actually for the teacher of ballet this is of very little importance, for the simple reason that ballet has its own technique of definite and exact movements. *If these movements are performed correctly the correct muscles will work.* (Sparger 1982: 10)

In Sweigardian terms, one would say, "If these movements are *visualized* correctly, the correct muscles will work."

I have introduced five issues regarding the teaching and learning of dancing. I will return to these issues in succeeding chapters, so we will repeat them here:

1. *Separation of (some) types of imagery.* That is, (a) images of locations of key points in the body and (b) aesthetic imagery. The images of bodily *locations* (item a) pertains to training and/or educating students' bodies. Aesthetic imagery (item b) pertains to how a danced movement is done to achieve the desired aesthetic or artistic effect.

   Teachers should think about separating the imagery pertaining specifically to the body in terms of *teaching it to perform or control the movements* of the idiom of dancing (for example, suitable for a *plié,* an *arabesque,* a Graham "fall," or a leg extension of some kind) and the imagery used to create (when performing) some desired aesthetic effect, such as "Your leg should be lighter," or, "Let your arms explode into the arabesque," or "The beating of the foot must be like a beating heart," and so on.

2. *Teaching dancers how to protect themselves from unsuitable bodily imagery,* and the question of how much (as traditionally taught) anatomy, kinesiology, and so forth is necessary for dancers to know in order to dance.
3. The question of how much anatomy, physiology, and so on is necessary for *teachers* of dancing, in contrast to physiotherapists, surgeons, and others in traditional health professions.
4. The question of how much actual anatomical knowledge is necessary for the *practice* of dancing, the *teaching* of dancing, and for *methods of retraining,* such as Ideokinesis, Feldenkrais, the Alexander method, and others.
5. The question of what the study of dancing entails. A French anthropologist, Michel Foucault, said that certain practices (including dancing) "effect by their own means or with the help of others a certain amount of operations on their own bodies and souls, thoughts, conduct, and way of being, so as to transform themselves in order to attain a certain state of happiness, purity, wisdom, perfection, or immortality" (Foucault 1988: 18).

Appendix to Chapter 1
Lexicons of the Body

Social lexicons of common English and French words for the body, in contrast to anatomical terminology, would look like this:

English social lexicon	French social lexicon	Anatomical terms
head	la tête	cranium
face	le visage	••••••
chin	le menton	mandible
throat (neck)	la gorge (le cou)	larynx/cervical vertebrae
shoulders	les épaules	scapulae
chest	la poitrine	thoracic cavity
breast	le sein	•••••••
waist	la taille	•••••••
hips	les hanches	pelvis
thigh	la cuisse	femur
leg	la jambe	tibia/fibula
knee	le genou	patella
calf	le millet	gastrocnemius
ankle	le cheville	•••••••
foot	le pied	tarsus
toes	l'orteil	metatarsus (phalanges)
hand	la main	••••••••
arm	le bras	humerus/radius/ulna
elbow	le coude	••••••••
wrist	le poignet	carpus
fingers	les doigts	metacarpus
abdomen	le ventre	••••••••
the back	le dos	"the spine"
"small" of the back	les reins	lumbar region

The above lists are taken from a two-part essay written in 1980 in the *Journal of the Anthropological Study of Human Movement* [JASHM], Vol. 1, Nos. 1 and 2, entitled "Taxonomies of the Body."

The main purpose of the articles (and this appendix) is to draw attention to the fact that *social* terms for the body (such as "hip," "thigh," etc.) are *general.* They bring to mind images of the *contours* or *outlines* of bodily parts, in contrast to anatomical terminology (such as "thigh joint," "iliac crest," "sacrum," etc.) that is *specific* with regard to the location of moving parts internal to the body.

The list of anatomical terminology as it is given is problematic, however, because it generally refers to *bone anatomy,* except for the words signifying "calf" in English and "*le millet*" in French. Here, we find the anatomical name for the *gastrocnemius* muscle in the lower leg.

It would be possible to construct a detailed list of anatomical terms for the social lexicon of the body using nothing but the names of muscles, but in Sweigardian terms, there would be no point in doing this because it encourages students to think that they can (or should) exert *voluntary control over specific muscles* that pertain to the movements they learn. There are so many muscles involved—even in simple movements—that it is impossible to determine which muscle to isolate. In any case, memorizing muscles (or their names) is neither useful nor efficient for learning or teaching dancing. Teaching should *never* include telling students what muscle (or muscles) to use to achieve any given movement. Ideokinetically, the best way to get the right muscles to move in a dance class is to use the centering properties of the light beams, outlined in Chapter 1.

Furthermore, the lists of English, French, and anatomical terminology emphasize the difficulty of teaching traditional anatomy to teachers of dancing. What, among the thousands of details of (even) bone anatomy, are they meant to use when they teach dancing? Sparger says it very well: "In presenting the study of anatomy to teachers of ballet, it is not easy to decide where to begin, and, more especially, how far to go" (Sparger 1982: 10—see page 16, Chapter 1, for more of the quotation).

The above list points to the difficulty of using the ordinary social lexicon of the body to teach dancing, and (for those who read Williams's work on taxonomies cited above), the anthropological reasons for learning how the body is named in each culture becomes fairly clear. Unfortunately, the subject of anthropological applications of bodily taxonomies is too complicated to address in this book, but some useful references are listed in Williams (1980) for those who want to pursue the subject further. Suffice to say here that *words commonly used for the body are, on the whole, too general to be effective* with regard to teaching movement.

Visualizing one's thigh or hip is quite different from locating the right or left ilio-femoral joint in the body. To see a mental picture of hip or leg means that students visualize the superficial characteristics of the bodily part. Ideokinetically, it is imagery that *locates the ilio-femoral joint* for each student that will ultimately improve the leg movements, because that joint and its surrounding musculature is where movements of the legs originate.

It needs to be said, however, that danced moves themselves should *not* be done visualizing the ilio-femoral joints (or any other joint). In my opinion, no dance form should be approached in this way. This is why my classes were divided into body techniques (for the purpose of teaching important anatomical *locational* facts to the dancers), and the actual teaching of the danced movement themselves, where the focus should be on each *whole movement* or movement combination that is required for the performance of the dance idiom.

Perhaps the point would be clearer by saying that a portion of each class should be devoted to functional anatomy and locations of key points in the body. The rest of the class should be devoted to the teaching of dancing.

Chapter 2
Relaxation

Throughout *A Kinesthetic Legacy,* Barbara Clark refers to the so-called relaxation features of ideokinesis as "table work" and "table teaching" (Matt 1993: 43–47, 49). I found her comments especially interesting because my introduction to ideokinesis was entirely through the constructive rest position (hereafter called CRP). Many sessions I had with Sweigard consisted solely of "table work" in the CRP.

Image 10. Throughout *A Kinesthetic Legacy,* Barbara Clark refers to the so-called relaxation features of ideokinesis as "table work" and "table teaching" (Matt 1993: 43–47, 49). I found her comments especially interesting because my introduction to ideokinesis was entirely through the constructive rest position (CRP). Many sessions I had with Sweigard consisted solely of "table work" in the CRP.

I was told that the CRP was a necessary beginning because I would not be able to change any of the bodily postures or habits I had *until I could successfully reduce the tensions already existing in my body* through using imagery given in the CRP. After these tensions were reduced, Dr. Sweigard could then set about establishing new patterns of neuromuscular response in my body (also using CRP) based upon the lines of movement (also called "lines of action" or "lines of force"). After I had improved to Sweigard's satisfaction with these lessons, she gradually introduced me to a side-lying position, then a sitting position, then standing, and finally, ordinary walking. In our private sessions, we always started with the CRP.

Clark's description of the CRP coincides with what I was taught; that is, "In constructive rest, the students reclined on their backs with their knees bent and oriented to [the] ceiling and their arms resting comfortably across the chest" (Matt 1993: 43); however, Sweigard told me that especially ballet dancers often must tie their knees together with a scarf (see Image 10) because their legs tend to fall open owing to habitual patterns of tension in the buttocks and outer leg muscles. Ideally, I was meant to understand that the knees should "prop each other up" in the CRP, which meant that *no muscular tension whatsoever* was required to hold them up.

Another adjustment that could be made in the CRP had to do with the upper arms. Because of habitual tensions in the shoulder girdle and upper back (a hyperextended chest), some students' arms would not rest easily across the chest. In these cases, the upper arms could be supported by pillows so that they were aligned with the coronal plane (divides body front to back) of the body. This meant that the hands rested on either side of the lower rib cage. Both of these adjustments (tying the knees together and support of arms, if necessary) maintained the notion that constructive rest was a position in which a skeleton could be balanced with no external aids whatsoever. In other words, the CRP itself *requires no muscular tension* to maintain.

Imagery in the CRP began with the suit of clothes (see Sweigard 1974: 232–36, "The Empty Suit") corresponding to the most superficial layer of muscles of the body. These images were emphasized by Sweigard's hands guiding the locations and directions of the images on my body. From there, as the sessions progressed, the imagery for tension reduction got deeper and deeper, until the final images I learned for tension reduction consisted of slowly removing knitted woolen covers from the bones themselves—all done, of course, with correct lines of direction and action in the body.

For example, when peeling knitted woolen covers off the *scapulae* (shoulder blades), one always started at the *medial* edge of the scapula (the edge nearest the spine) and worked toward the outer edge and the gleno-humeral joint. None of the images she gave me were ever to be changed with regard to the *direction of the image*. That is, the back of the suit of clothes was always to be smoothed from the shoulder seams *downward* toward the hem of the coat—never in the opposite direction. The suit coat was to be smoothed *from* the center back seam *toward* the front zipper that closed the coat—never from front to back and so on.

Ordinary Relaxation versus Tension Reduction

Dr. Sweigard emphasized many times and in many ways the difference between "relaxation" as it is ordinarily conceived, and the tension reduction she tried to accomplish as a result of her lessons. She described ordinary relaxation techniques as "undisciplined, random affairs," such as people might use before going to sleep at night, or the relatively relaxed state of muscles when lying on a couch or in a reclining chair. According to her, tension reduction was not the same because it was defined by *purpose* and the student's mental control over the specific imagery involved. Moreover, tension reduction was to be practiced in the CRP.

Sweigard taught that the ideokinetic imagery for tension reduction was never to be confused with images for ordinary relaxation. For example, she pointed out how a person might (prior to going to sleep) start with the feet by creating a mental picture of the feet, then "telling" the feet (or some other bodily part) to relax. While these sorts of things were good sleep aids, helping to lessen fatigue, they did not produce the same results as tension-reduction imagery, both because they rarely went deep enough to assist the body in changing muscular habits over time, and because the process frequently ended in sleep.

One was meant to get up from tension-reduction sessions mentally more alert and refreshed than when one began the work. If the CRP tension-reduction sessions ended in sleep, the student was doing something wrong—sleep is the result of either lack of focus or an undisciplined mind, or both. Furthermore, if students found they had problems with mental *activity* after completing exercises in the CRP, Sweigard's advice was to know how to direct one's thinking. She was convinced that "Change in muscle action toward greater efficiency requires directed mental activity in watching specific action in a *concentrated* manner" (1946: 22). See the Appendix at the end of this chapter (page 35) for the entire text of her article.

> Try the following for the next month. Take the constructive rest position. Then consider yourself a bag of sand. Your work is done; from this point on until you get up from the floor, you do nothing but watch intently the flow of fine, dry

> sand. Watch sand flow from the knee, down the thigh into the pelvis. Then from the knee down the lower leg and out through the base of the heel. The covering collapses together in each area as sand flows away. Now the other leg. (1946: 21)

I have included Sweigard's writing above and in the Appendix because I do not want readers to depend solely upon my interpretation of what she said.

It is informative, I think, to contrast Sweigard's ideas with the role of relaxation in the "rest and activity phases" in a kind of total bodily economy and balance that Mabel Todd advocated:

> Complete relaxation would be death. In complete balance the human structure would be devoid of activity. Living, we shall never attain either. But unless we struggle toward these ideals, balance and rest, and acquire a knowledge of principles underlying the balance of bodily materials and forces, there is no way of dealing intelligently with the extreme imbalances, emotional and physical, of living.
>
> Our object is that of preserving that fine bodily "instability" . . . where available energy hangs in the balance, ready to be converted at an instant into its working form, movement. (Todd 1937: 262*ff*)

Todd's chapter (1937: 262–95) is instructive, both because it offers a picture of fatigue in minute detail with reference to several bodily systems (muscles, nerves, organs, etc.) and because Todd clearly defines what she means by *relaxation*:

> Relaxation is the equilibrium of the rest and activity phases in the rhythms of living, rest being the passive phase between the active phases of these rhythms, as for example, in the rhythms of the heart, diaphragm and reciprocal muscle action.
>
> Relaxation is potential balance, experienced in the ideal state of well-being.
>
> *Relaxation is not negation, it is not passivity.* The moment it is considered as such, flaccidity is encouraged, both mental and physical. Take hold of your bones *softly*, but do not let go of them. (Todd 1937: 293)

Even though I encourage students to read Todd's book, I do so with reservations, some of which may be the same as those held by Sweigard, who said:

> Having been a student of Miss Todd it is often assumed that I continue her method of teaching. This is true only in part. I accept her basic philosophy, but my teaching has its basis in the findings of two of my research studies and, therefore differs in approach quite markedly in many ways from that of Miss Todd or any of her pupils. (Matt 1993: 43, from a letter Sweigard wrote to Matt, March 16, 1972)

Much later on, I found that the practice of Sweigard's tension reduction was difficult for a few students for several reasons: 1. some were unable adequately to concentrate; finding that they had to increase their ability to focus their attention for long periods of time; 2. a few had habits of tensing certain muscle groups (for example, the neck, leg, or facial muscles) when required to concentrate, although they were usually unaware of the fact that their bodies automatically responded in this way. To them, concentration (mental focus) was consistently associated with muscular tensions in some bodily part(s). This kind of habitual response is hard (though not impossible) to change. Furthermore, I was instructed by Sweigard not to try to carry over any tension-reduction images used in the CRP into an actual class, rehearsal, or performance of dancing. I took her admonitions and comments seriously.

Utilizing Imagery in Dance Performances

Matt points out that "Barbara [Clark] realized that dancers, already being quite kinesthetically aware, could readily utilize what she was developing to enhance dance performance and avoid injuries" (Matt 1993: 43–44), and "Barbara's approach was certainly simpler and more casual than Dr. Sweigard's" (*Ibid.* 44). I begin with Clark's comments, not to criticize what she said or practiced, but to emphasize a major difference in the two women's usage of imagery. Since I never saw, or participated in, any group lessons that Sweigard taught, my impression of Sweigard's work was that it led only *obliquely* to the improvement of dance (or any kind of movement) performance.

She never instructed me to use ideokinetic imagery in any movement *activity* other than walking and the small moves that she describes in her chapter on "Voluntary Movement" (Sweigard 1974: 274–306). From my experience, I concluded that what ideokinesis was about was retraining the habits of my (or anyone's) body so that they functioned more effectively in *any activity whatsoever,* including dancing. It may be the case that other of Sweigard's students had their training carried over into the performance of movement (danced or otherwise), but if so, I know nothing about such training, nor did I experience any such thing with her.

I believe that Dr. Sweigard's extraordinary grasp of anatomical and physiological facts about the body led her to restrict ideokinetic training to the *reeducation of neuromuscular patterns* of the students' bodies such that all their "performances" (everyday and otherwise) were improved. It was my impression that she really wasn't interested in anything else, thus I am not sure what Matt means when she says,

> with all of her scientific background, the theoretical underpinnings of the teaching were uppermost in Dr. Sweigard's thinking. Working in an academic

> environment, it was important to supply the students with the logic and references they would need to understand the material intellectually. (Matt 1993: 44)

Maybe the answer lies in a simple point of comparison, i.e. "Barbara avoided intellectualizing the work, feeling that too much attention to the rationale ultimately detracted from the student's creative exploration of the imagery" (*Ibid.* 44). My interpretation of the phrase is that Clark's interest was focused on kinesthetic awareness and the science of *movement,* rather than Sweigard's "postural imbalances" and "ideokinetic *facilitation.*" That is,

> **Ideo,** the idea, the sole stimulator in the process, is defined as a concept developed through empirical mental processes. The idea, the concept of movement, is the voluntary act and the sole voluntary component of all movement. Any further voluntary control only interferes with the process of movement and inhibits rather than promotes efficient performance. **Imagined movement** is best defined as an ideokinetic facilitator. (Sweigard 1974: 7—bold typeface added)

In contrast, according to Clark, "We should not allow the subject matter to eclipse the above purpose [teaching kinesthetic awareness]. The *science of movement* is mechanical, and does not necessarily include kinesthetic awareness" (Matt, citing Clark from her diaries, 1993: 45—italics added). Clark was convinced that "The object of the work is to teach kinesthetic awareness; everything else must be subordinated to that" (*Ibid.* 45—from an entry in Clark's diary). I therefore agree with Matt that there was a "dissimilarity in philosophy which was responsible for the contrasting approaches of the two teachers toward the use of imagery" (*Ibid.* 45).

I did not know Barbara Clark, but even if I had, we would probably have arrived at different conclusions in terms of approach to the study of ideokinesis, mainly because I never carried Sweigard's imagery over into the creation of danced or theatrical movement. Given the training I had, I was prevented from doing so, because thinking—the ideas—*without any muscular activity whatsoever* is the basis upon which Sweigard's ideokinetic practices rested. Matt cites Sweigard thus:

> You know the one marked feeling I have had about all those who try to carry on in some manner the work Miss Todd started—a lack of scientific background and hence inability to stand up to anyone in supporting the manner of the work. One wins by changing the person, but there is no winning by writing and by talking unless one can do it with readiness to cite facts and principles. Miss Todd had them in part, but still not enough; she lacked true research backing and was unable to think in those terms even after such research had been done. She truly abhorred the scientific attack as she let me know when we were working together. (Matt 1993: 68, from a letter to Barbara Clark, 23 January, 1957)

I am convinced that the greatest value in Mabel Todd's work is that she made *the initial connection* between imagery and *the living body.* This was a remarkable insight, given the traditional (and centuries long) split in western thinking between mind and body, owing to the legacy in the Western world of René Descartes' philosophy. Todd's achievement was extraordinary. She did not, however, confine her accomplishment to bodily posture alone *as the basis for* any movement whatsoever, as Sweigard did.

To my way of thinking, the imagery and movement connection is a misconception of what *Sweigardian ideokinesis* is all about. That's not to say there is *no* relationship between imagery (*writ large*) and movement. Certainly, there are such connections and I will be specific about the subject in later chapters, but here, I aim to discuss ideokinesis in terms of Sweigard's approach to bodily training and reeducation, which, as far as I know, did not include training in any movement or danced form except ordinary walking and the small voluntary moves described in her book.

As I tried to point out in her memorial tribute (New York University, 1982), Sweigard came to these conclusions: "(a) posture is an expression of habits of body mechanics, (b) balance and movement of the body take place by means of conditioned neuromuscular reflex action without so-called voluntary control of the individual, (c) the human body functions as an organismic whole through the integrative action of the nervous systems, (d) principles of mechanics apply to the human body as to inanimate structures, (e) thinking influences muscle action, and (f) changing habits of neuro-muscular responses to stimuli is basically an *educational* procedure, not an *exercise* procedure" (Sweigard 1939: Preface). In the memorial tribute to her, I also pointed out that

> She was afraid that her work would be misused and the result of her refusal to codify her work has been much of what she taught single students that was meant to be applied only to specific problems is now being taught as the whole of ideokinesis. The eagerness with which some have attempted to disseminate the undoubtedly beneficial effects of her work and the novelty of her approach has generated numerous misconceptions. Nearly everyone can benefit from the study of ideokinesis, but comprehension of it *as education* requires a sophisticated mind and equally sophisticated knowledge of several areas of education. Her work is too often seen merely as a 'treatment' or a 'cure,' and is wrongly associated with a mystical 'laying on of hands,' even with forms of hypnosis. (Williams 1982: 9–10)

In other words, ideokinesis exists in a larger world, as Todd found out to her cost: "In the early 1950s a lawsuit was brought against Mabel Todd by a

New York medical association accusing her of practicing medicine without a license. Although the suit was not brought to trial, Barbara learned Miss Todd agreed to terminate her work in the state of New York as part of an out-of-court settlement. Todd closed her studio in the Roosevelt Hotel and dismissed Ruth Mitchell, at the time her only assistant. Although Todd's work on *The Hidden You* and her failing health had already reduced her influence in the teaching [of neuromuscular reeducation], the news of the scandal was a tremendous setback" (Matt 1993: 50).

I feel sure that one of the reasons why Sweigard was adamant that students understand the rationale of ideokinesis was that she wanted them to understand that ideokinesis *wasn't medical treatment, but movement education*—and with good reason. To my certain knowledge, she felt the same about other areas of human endeavor: spiritual, the mystical (or occult), the artistic, and the psychological.

Spin-Offs

Bob Fleshman rightly ascribes the discovery and development of ideokinesis to Sweigard (1984), but he did not study with her. He studied with me. He once said that my work in the dance enabled him to "crack the Decroux code of movement" and the art of mime. Apparently, developments with reference to imagery and movement also occurred through Ahsen's work on "Imagery, Drama and Transformation" (1984). These will not be discussed further here, but for serious students, Ahsen's work will be of interest.

In general, I think Fleshman's work with mime students coincided *in some ways* with Sweigardian ideokinesis; however, he departed from what he was taught when he attempted to *superimpose* Sweigard's eight lines of movement (which he calls "lines of energy") over the upright lines of Decroux's "Zero" position, saying that

> At a certain point the student has accomplished enough in the floor work of Ideokinesis and in the training exercises of Decroux that the two systems can be combined. The student is brought from the floor to an upright position. The transition of image work and body control from the horizontal position to the vertical one is somewhat difficult and rather tricky. The eight lines of energy of Ideokinesis are established on the vertical, superimposed with the upright lines of Decroux's "Zero" or "Neutral" [position]. Especially important is the long line of energy running lengthwise through the middle of the body, at center. In an upright position this is the line of gravity, where the weight of the body enters the earth below. This is the center line for mime, and in my work it is usually referred to as the "silver rod." It is the point of reference for movement,

> and the main and central image. *There are some minor adjustments that must be made between the alignment of ideokinesis and [the] artistic neutral of Decroux's mime, mainly placement of the weight (center of gravity) in a much more forward position than in ideokinesis.* (Fleshman 1984: 96–97—italics added)

The "minor adjustments that must be made" between ideokinetic alignment and the placement of weight in Decroux's mime is (or should be) a major topic of concern to anyone who is interested in teaching Sweigardian ideokinesis to people in other movement disciplines. The topic has to do with (in Decroux's mime) placing the weight in a standing position "*in a much more forward position* than in ideokinesis." I draw attention to Fleshman's "alignment" example, because it is just here that I believe the demands of a movement idiom and ideokinesis (in a standing and/or walking position) are radically different, as we shall soon see from an example drawn from east Indian dancing.

Sweigard's postural alignment (the axis of gravity for the torso in a standing position) would have to be altered for Kathak (a north Indian style of dancing), as well, because of the "slap" of the feet connected with all footwork of this idiom of dancing. The axis of gravity in Kathak has to be farther *back* than it occurs in a standing or walking position in ideokinesis. "Leaning" the whole body forward (the requirement for mime) in a standing or walking position, the muscular pulls throughout the entire skeletal structure change. The same is true for the initial standing position of Kathak, where the body leans slightly back. In either case, the body has to some extent be "held" so that it doesn't topple over.

Neither alignment is, therefore, a bodily position that follows one of the basic rules of ideokinesis: "all voluntary contribution to a movement must be reduced to a minimum to lessen interference by established neuromuscular habits" (Sweigard 1974: 6). It is, I think, relevant to read what Sweigard says:

> The method presented here emphasizes its neural aspects; these hold the secret for securing and maintaining neuromuscular efficiency, with all its salutary benefits. In this respect, the method differs significantly from the procedures generally employed in teaching posture, fitness, and therapeutic exercises. In fact, it differs from the general procedure for teaching skills—be they in an occupation, a sport, or one of the performing arts. Although the standard methods often produce the outward effects desired, they frequently build strain which can, in turn, lead to premature debilitation and actual loss of efficiency in movement. (Sweigard 1974: v)

The axis of gravity of the whole body in space is a concept that, to my knowledge, didn't (and doesn't) exist in Sweigardian ideokinesis because

Image 11. Axis of gravity in "natural" position.

Image 12. The "zero position" of Decroux's mime; axis of gravity is forward.

Image 13. Beginning posture for Kathak; axis of gravity is back.

the axis of gravity as it is conceived by mimes or dancers* falls *between* the feet. It is an imagined line that passes through the head and the torso, and between the legs to the floor. Sweigard's use of imagery in the standing position is illustrated below:

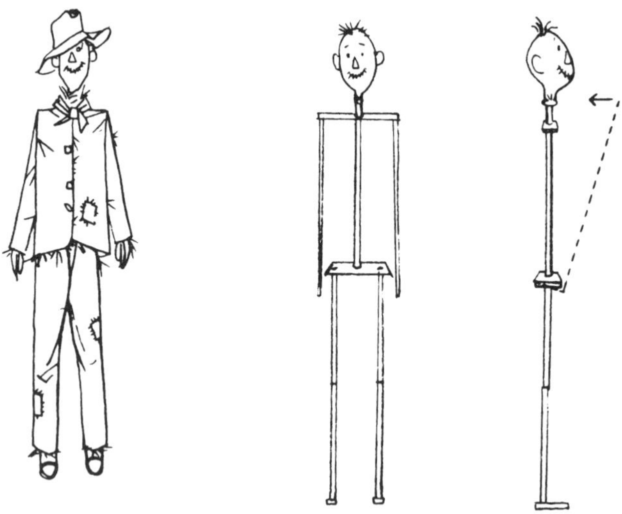

*The axis of gravity in the ballet is often referred to as "the plumb line" and is directly between the heels (at the ankle joints) of the standing dancer.

Notice that there is no line that extends from the torso to ground level. The axis of gravity is only in the trunk. Also notice that in her use of the empty suit, Sweigard translates it to the image of a scarecrow, because "you, as the empty suit, need inside supports similar to the scarecrow [left]" (1974: 236). She goes on to say,

> In the pelvis there is a level board extending crosswise, half way from front to back, to support the vertical rod for the head and a crosswise rod for the shoulders of the coat [center]. Unfortunately this board tends to tip downward in front in everybody, and this tipping means that the top of the vertical rod which it supports will be far out in front of the base of the head, not under it [right]. In your mind's eye watch the crosswise board moving upward in front to place the board in a level position (do not help), and bring the top of the vertical rod back under the center of the head. The crosswise board [pelvic level] is supported in turn by a vertical rod under each end. These rods extend to the ground through the trouser legs. In your imagination pay attention to the inner supports; then use any imagined movements in the empty suit that appeal to you.
>
> As you walk while being an empty suit pay no attention to the trouser legs; concentrate on movement in other parts of the suit (your trunk and head) and walk *naturally.* (Sweigard 1974: 236)

Neither mimes nor Kathak dancers, nor ballet or modern concert dancers, can be said to walk "naturally" from an ideokinetic point of view.

Keen students will perceive that, in giving images to lengthen the spine downward, not all of the images Sweigard uses are for a standing position (1974: 237–39). Later on, she says,

> Many images for muscular re-coordination to effect skeletal realignment can be practiced while sitting, standing, or walking. The maximal benefit from any image is attained only if there is a full awareness of the ideal skeletal alignment and a clear concept of the problems of balance in these positions. The role of imagined movement, then, is not only to produce better alignment, but also to provide increasingly greater stability in the improved alignment. *Stability* is invariably a function of continued practice to put firmness into the newly evolved neuromuscular patterns of work. (*Ibid.* 251)

"Ideal skeletal alignment" is not achieved by leaning the axis forward (as in mime), or by leaning slightly back (as in Kathak). Nor is it achieved by any alteration whatsoever in the notions of "ideal skeletal alignment and a clear concept of the problems of balance" in the standing position, which in ideokinesis is designed for balanced, *everyday* walking and standing. It is well to remember, too, that in ballet dancing (and in most modern concert idioms of dancing), the standing position is connected with a "turn-out" of the legs—again, a departure from an ordinary standing position.

Consider, too, that the positions of arms in danced and mimed movements are "held." That is, they are not "natural positions" of arms, just as the feet, especially in the ballet, are altered. A Martha Graham contraction and release is not a "natural" position of the body, but it is not, I hope, necessary to enumerate all of the positions assumed by trained dancers in all forms of dancing to point out that none of them are "natural" in the sense that Sweigard used the word.

Having said that, perhaps I should add that ideokinetic imagery should be left out of mime and dance forms altogether. The imagery that is used in a dance or mime class should pertain to the *locations and directions* of the danced or mimed moves in terms of the idiom itself, for, as Sparger so aptly put it, "ballet has its own technique of definite and exact movements. *If these movements are performed correctly the correct muscles will work*" (Sparger 1982: 10).

Not only has the ballet its own technique of definite and exact movements, every form of dancing (and mime) has its own technique of definite and exact movements. *If these are visualized accurately,* the body will perform them as well as it can. It will perform them even better if the performer has a basis of ideokinetic study that has helped him or her to improve the efficiency of the neuromuscular coordination engaged habitually in maintaining upright balance (Sweigard 1974: 174).

While on the subject of Sweigard's imagery, it should be said that she uses very little anatomical imagery in her published work. In her work with me, she used many images of bones and their relationships, but she did so, I think, because she insisted that I take a class in bone anatomy before studying with her. Furthermore, all the sessions I had with her involved what I think of as "tactile imagery," as well. That is, she located every image and its direction in my body with her hands in the CRP and in sitting and standing positions. There was never any doubt about (a) *what* the image was, (b) *where* it was located with reference to my body, or (c) *what function* it was meant to complete with regard to changing habitual postural patterns. There is a sense in which this training explains why I used tactile pressure points in teaching the light beams (see Chapter 1), either by having the students work as buddies or having them use their own fingers to locate various places in their bodies.

And there is another reason: students sometimes develop the erroneous notion that they can "do" images voluntarily. They come to believe (through some mysterious process of self-persuasion) that if they picture the thigh joint while moving one of their legs, that this is what ideokinetic imaging amounts to. It isn't. This is the reason why I separated a portion of the class at the beginning for the teaching of body techniques from the actual dancing class itself. This strategy places emphasis on the fact that work on students' bodies should be kept separate from whatever activity they did. This lack of separation forms the basic criticism I have of Eric Franklin's book (1996). Franklin produces an overabundance of images for the body, not telling us *when* these images are to be used or *why* they are used—only *that* they should be used.

For now, however, I would like to direct the reader's attention to Sweigard's chapter on voluntary movement where she says:

> 7. *Voluntary movement must be slow* in order to allow time for imagined action to exert its influence on the subcortical patterning of the work of agonists, antagonists, synergists, and stabilizers. The faster the movement the greater the probability that it will be performed by the very neuromuscular habits one is trying to change toward increased efficiency.
>
> 8. *Imagined action during voluntary movement of any part or parts of the body is invariably located in the trunk.* Its direction is often in the opposite direction of the distortion the voluntary movement will produce in the trunk. (1974: 276—italics added)

Notice that she gives emphasis to *slowness* of movement. Dance classes are rarely slow. It requires a different orientation to the teaching of movement to accomplish what Sweigard says in item 7. Moreover, in item 5 on page 276 in her book, the need for *complete rest* between each attempt to perform her prescribed voluntary movements is highlighted. She also says,

> 6. Imagined movement should be used both before and during voluntary movement. *There is always a tendency to start the actual movement too soon,* once it is understood, without taking time to concentrate first on the imagery being used. The sequence after the movement has been learned is (*a*) complete rest, (*b*) imagined action, (*c*) slow movement with continued *concentration on imagery,* (*d*) sudden release of effort, and (*e*) sensing the change, if any. (*Ibid.* 276—italics added)

The point for dancers is this: can dance teachers (a) reduce the need for endless repetitions of "exercises," inside or outside dance classes, and (b) significantly eliminate pain and suffering, such as Gelsey Kirkland describes?

> Balanchine and his teachers unwittingly encourage young dancers to self-destruct, rationalized as part of the sacrifice that must be made to the art. The speed and shortcuts that he built into the training process called for physical

> cheating in which the dancer distorted the body to deliver the position or step that Balanchine demanded. The risk of injury was ignored. I watched many of my friends become casualties and fall by the wayside. (Kirkland 1987: 28–29)

In other words, although "practice does (sometimes) make perfect," it can (and commonly does) make perfect mistakes. Had Gelsey Kirkland had ideokinetic training, I think she would have avoided the various traps she describes, for she would have learned how to reduce tension in her body and how better to relate images to movement, as we shall see more clearly in Chapter 4.

To Sweigard, better dancing could only be achieved through the development of better posture. Better posture could only be achieved through changing established neuromuscular habit patterns in the body. In turn, better habit patterns in the body can only be achieved through the discipline of constructive rest.

We will now turn to the notion of kinesthesia and "kinesthetic awareness," traveling a circuitous route through Chapter 3 because of the relationship between danced movements and the "everyday moves" in which all dancers (and everyone else) participates. However, since this book is primarily about Sweigard's methods of teaching, I will highlight one of her methods as I experienced it by talking about baking biscuits.

I will conclude by pointing out that Dr. Sweigard moved superbly well, with great economy of effort, controlled energy, and focus. She was a living example of the postural integrity and movement efficiency that she advocated, and she was always a living image of the realizable *potential* that one could achieve in ordinary everyday movement.

Appendix to Chapter 2
Constructive Rest

LULU E. SWEIGARD

Consistent daily rest in the constructive rest position produces various changes in the body. The pressure area of the pelvis on the floor becomes greater. Any discomfort from pressure at the end of the spine (coccyx) lessens or disappears. The low back rests closer to the floor; its hollow is not as great. The knees stand upright more easily. The arms should lie across the chest more comfortably. However, continued tightness in the muscles between the shoulder blades and spine may still pull the arms off the chest unless they are held there. Do not hold! In the rest position, holding any part of the body in a specific position is not rest, it is added muscle work.

"Kinesthetic" awareness increases with daily rest. That is, one becomes more conscious of what is happening in one's body; where muscles are tight, strained, or even painful; where and when muscles relax and become more comfortable; where the body contacts and presses on the floor, and the degree of this pressure.

The next step toward a better posture is learning to direct one's mental activity during rest. Each person is aware of the way he can "tie himself in knots" with mental activity and emotional strain. This is felt especially in the shoulders and back of the neck. Such is almost everyone's personal experience with the fact that *thinking and emotions influence muscle action.* If thinking can affect one's body detrimentally, it can also affect it advantageously. The important thing is to know how to direct one's thinking. *Watching* activity in the body, just as if it was really taking place, can change the manner in which muscles are working. What action to watch, its location, and its direction are most important.

Try the following for the next month. Take the constructive rest position. Then consider yourself a bag of sand. Your work is done; from this point on until you get up from the floor, you do nothing but watch intently the flow of fine, dry sand. Watch sand flow from the knee, down the thigh, into the pelvis. Then from the knee down the lower leg and out through the base of the heel. The covering collapses together in each area as sand flows away. Now the other leg. If the knees tend to fall apart,

---

Source: *Self Magazine* 1(4), June 1946.

pigeon-toe a bit more, or tie the knees to each other if necessary. If the feet tend to slide away from your body, make the roll of rug under the balls of the feet bigger.

Next, watch sand settle to the floor throughout the trunk. In the pelvis, watch it spread sideways on the floor. In the arms, watch sand flow from the elbow downward through the shoulder into the main bag of sand, the trunk. Then watch sand flow from the elbow down the lower arm and out through the fingertips. The sand spreads, sideways on the floor, in the region of the shoulders. Lastly, visualize your head as completely empty.

It will take twenty to thirty minutes to watch all the movement of sand. Start with the legs; go back to them later. Review the procedure, in any order you wish. When your "watching" work is completed, you see yourself somewhat as a rag doll, without stuffing. Practice just before going to bed. If possible, practice several times through the day to speed change in your body.

When sitting, standing, or walking, experiment with watching sand flow downward—always downward—in different parts of the body. In the upright position, one needs to visualize some inside supports such as those used in a scarecrow—a central vertical rod in the trunk, on which the empty head sits; one in each leg; and a crosswise rod in the shoulders. In the mind's eye these supports do not bend, no matter how much the covering of the sand slumps inward and *downward* as the sand flows.

Do not give any voluntary help to the flow of sand at any time. This is the most common mistake made in the practice of this procedure. Do not rely on "wishful thinking." It will not produce results. Change in muscle action toward greater efficiency requires directed mental activity in watching specific action in a *concentrated* manner. This is a discipline, but it is also a means of thwarting the many harmful effects of the strain of modern living; it is a start toward a better posture.

Chapter 3
Baking Biscuits and Kinesthesia

We start with the idea of an interface between ideokinesis and teaching dancing. For me, the diagram illustrates a personal choice:

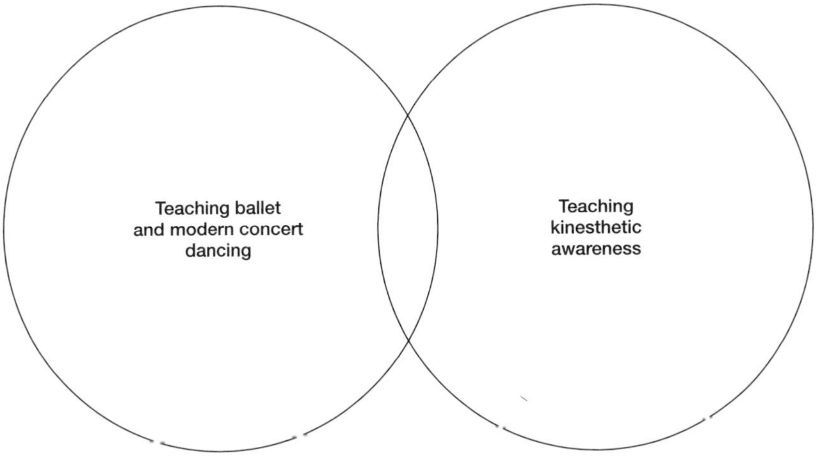

Drawing 1. A Venn diagram conception of choice. There is an *overlap* that connects different fields of work, but the overlap *is not the same as* either of the fields of work concerned.

More important to realize, perhaps, is the fact that there are huge differences between teaching ideokinesis to nondancers and dancers. The area of overlap grows smaller for dancers because the aims and attitudes of serious dancers toward their bodies are in a class by themselves. On the whole, professional dancers are in a distinctly different social group that is defined by what they do. To a professional dancer, the body is an *instrument* or vehicle through which he or she produces the performances required to serve the repertoires of professional companies.

Depending upon the idiom of dancing, ballet or modern concert dancers have developed higher ranges of flexibility of their joints and maximum amounts of strength, speed, and balance required to serve the needs of their chosen dance form. They do so in specific ways; for example, see Hart-Johnson's account of a Graham technique class (1997).

Nondancers do not place comparable kinds of long-term, persistent demands upon their bodies that are common among professional dancers, champion ice-skaters, or Olympic gymnasts. I think it safe to say that the majority of nondancers simply want their bodies, movement-wise, to function normally, *without pain,* so that they can participate in everyday life easily and freely. In my view, Dr. Sweigard carried the notion of everyday movement to the level of an art, although she would never have admitted this, nor, I think, did she see herself doing anything of that kind.

I often remember one of the times when I went to her house for a lesson, arriving to find her in her kitchen baking biscuits. The moves she used to accomplish this task became part of the lesson for that day. I watched her knead and roll the biscuit dough, reach for the knob on the electric stove that controlled the oven's heat, and, later, kneel as she opened the oven door to place the unbaked biscuits into the oven. She stood by the kitchen sink, washing the bowls, utensils, and measuring cups she had used. Then, she removed her apron and sat in a kitchen chair talking to me while we waited for the biscuits to bake. The telephone rang during this time, so I saw her rise from her chair, walk to the telephone, answer it, and return to her chair. When the biscuits were done, we enjoyed them with butter and jam while the discussion of functional anatomy and body mechanics continued.

Every move and postural detail imaginable connected with baking biscuits was noted by me and explained by her during our conversation over the kitchen table. I cannot overstress the fact that *every* move, every transition, every posture was gone through with the proverbial fine-tooth comb. Needless to say, perhaps, all moves that were required for the task were performed beautifully, although it would have embarrassed Dr. Sweigard if she had heard me say so, and she would have in no way conceived of what she did as a performance. Performing was for dancers, actors, and such—not for her!

Dr. Sweigard moved superbly well, with more controlled energy and focus than I have ever seen before or since. She practiced everything she preached. She personified the kind of postural integrity and movement efficiency that she advocated. She was a living image of the realizable *potential* that one could achieve in ordinary, everyday movement. She had to be at least seventy years old when I knew her, but her posture and movements were those of a slender twenty-five-year-old woman, although she had gray hair and a few extra lines on her face.

I never doubted that she could have carried the movement she did into artistic forms of mime or dancing, but she was shocked at my once-only tentative suggestion about the subject. She was, in her words, "just an ordinary woman who knows the potential her body has for moving (scientifically) correctly and well." She was that kind of down-to-earth, "all wool and a yard-wide" person. Her talk (lectures, discussions, or whatever you want to call them) was as direct, forthright, and economical as her movements were.

I believe that she finally wrote her book about ideokinesis when she became aware that she hadn't much longer to live, because she said several times in passing during the sessions we shared that she felt she didn't know enough to write a book! I later incorporated the everyday movement aspects of her teaching into classes I taught for the University of Wisconsin Extension Division in Green Bay, Wisconsin.

Initially, there were a group of a dozen women who enrolled in a beginning modern dance class when I commenced teaching in Green Bay in September 1963. At first, I *did* teach them some moves from modern dancing; however, as the average age of the class was thirty-five, I didn't have much scope for teaching them modern dance techniques, so, after warming up we got into ordinary, everyday movement and the mechanics thereof. As most of them were housewives, secretaries, or nurses, I drew heavily from their own experiences with movement, starting with the scene I described from Sweigard's house.

After constructing an imaginary kitchen (using some real props like tables and chairs), we had a great time going through these movements, and at their suggestion, we tackled others, such as cleaning the living room; scrubbing the basement floor; vacuuming; and taking down, laundering, and returning curtains to their rods. Some of them had babies, so we went through numerous movements connected with caring for babies. The nurses were interested in the mechanics of making beds and other tasks they had to perform at the hospitals where they worked.

We did all of these things in such a way that we pinpointed the bodily tensions they carried throughout their day. I taught them a few things about reducing tensions in their bodies, but for this I depended not on Sweigard's work but on my own version of the relaxation exercises that I had learned in hatha yoga classes with Swami Vishnudevananda in New York. They were not, however, permitted to carry their relaxation exercises into a state of sleep, following Sweigard's differentiation between tension reduction and ordinary relaxation. I didn't use her imagery—that is, the suit of clothes, the bag of sand, and so on.

The Green Bay group learned to sit down on chairs, sofas, and stools gracefully, not in some silly way, based on misconceptions of good etiquette, but through learning the mechanics of these everyday moves. They learned about what they were doing incorrectly that caused some of them to have headaches or backaches, and/or generally to feel worn out at the end of the day.

We acted out excursions to the grocery store, learning how to push shopping carts, how to get in and out of motorcars, how to carry babies and bags of groceries without throwing themselves out of balance or adding to the postural problems they already had. These women were students of mine for as long as I taught for University Extension Division. The original class developed into several classes structured along the same lines. And we didn't stop with everyday movement. We *did* get into dancing.

Mainly, I taught several of these sessions using various idioms of ballroom dancing, working up to modern concert dancing to the extent they were able, but by the time we had worked on *their* basics in everyday movement, they were equipped to acquire a real appreciation of what professional performers in highly sophisticated dance idioms achieved, and how some of that related to the ways in which they moved.

They developed different perceptions of movement in general—their own moves and those of professional dancers. They began to be aware of the differences in joint flexibility, strength, and control of professionals and themselves and much else. Through this approach, they became *intelligent spectators* of dance forms that they would never be able to perform. This is all part of the overlap (see Drawing 1) to which I referred at the beginning of this chapter.

The Green Bay students acquired enough knowledge to be sensible parents to their children who wanted to take dance classes. They learned enough about imagery in typical dance classes to be able to judge for themselves whether a dance teacher knew (or did not know) what he or she was doing. They learned how to ask questions, and without exception, they reported good results in their lives.

That is, many of their husbands, families, and women-, and men-friends noticed improvements in how they moved; they became better-than-average ballroom dance partners; and they developed high degrees of kinesthetic awareness. Some of them asked me about everyday movement routines that bothered their husbands (painting houses, lifting heavy cartons, and such), because men, of course, have similar problems with body mechanics. They, too, experience pain and suffering, although their skeletal structures are different in important ways.

The Green Bay students told me that, in general, their lives weren't boring anymore because they no longer had the same attitudes and ideas about the moves they did every day. Their lives were greatly enriched because they learned to understand their movement potential in the contexts in which they lived. Briefly, *they discovered kinesthesia,* described by an anthropologist of human movement, Dr. Brenda Farnell, thus:

> Kinesthesia [is] our sensory awareness of the position and movement of the body. I ask the reader to please close your eyes and lift your arm—move it around and ask yourself how you know where your arm is located? This is kinesthesia; literally 'movement' (kinetic) + 'sensitivity' (aesthesia). It is this kinesthetic sense that provides information on the whole repertory of our motor actions, from the raising of an arm, to walking, even to the turn of the eyeballs and swallowing. Physiologically speaking (that is, in the discourse of Western natural sciences), kinesthetic sensations are registered by receptors in the muscles, tendons and joints of the body. As the muscles function when we move bodily parts, various patterns of pressures on these receptors provide essential information for the guiding of motor action.
>
> The perception of spatial movement and orientation of the body as a whole also involves a fluid-filled receptor system located in the vestibules of the inner ear. More than balance, this structure provides the means by which we are aware of being tilted, shaken, or whirled about, and how, most of the time, we know "which way is up"!
>
> The exclusion of kinesthesia from the Western taxonomy of the senses—this (ab)sense, as it were—is particularly interesting because scholars of perception as diverse as Descartes, Dewey, Gibson and Merleau-Ponty all acknowledge body movement as the unexamined ground of all sensory perception. One is led to ask why, then, has kinesthesia been excluded from consideration? (Farnell 2003: 133–34)

It has always been interesting to me that the kinds of discovery made by my Green Bay students were discoveries about the very sense that, on the whole, has been *excluded* from the Western taxonomy of the senses. Perhaps kinesthesia is excluded because we are taught to think of human beings having *five* senses (sight, hearing, touch, smell, and taste), not six. Apparently, the sense of touch is thought to include the whole of kinesthesia, although it obviously doesn't.

It does not include the *spatial* characteristics of kinesthesia, for a start, which with regard to dancing is of fundamental importance (see Williams 2000 for more discussion). In my view, the spatial aspects of kinesthesia are just as important in daily life, and there is no reason why people should not enjoy the same greater awareness of themselves and moving in ordinary life as dancers enjoy on stage.

Furthermore, I want to say that I don't think the sciences of kinesiology, physiology, and related fields have done much to remedy this "unexamined ground of all sensory perception," as Farnell puts it, partly because few kinesiologists, physiologists, and biologists saw the need for a remedy in the sixties and seventies. These disciplines (including biomechanics) define themselves legitimately as sciences, but the boundaries of their disciplines do not include anything about the social, linguistic, and cultural aspects of movement; however, this is what the anthropology of human movement is all about.

> Great insights into the anatomical mechanisms of the human body have been achieved by Sweigard's kind of research, but insight into the nature of danced actions themselves are not achieved through anatomical, physiological or kinesiological study. The reason is simple: these sciences are not concerned with the social identities of dancers or signers, nor are they concerned with the meanings of dances, sign languages, rituals, and such, because "the meanings of perceivable actions involve complex intersections of personal and cultural values, beliefs and intentions, as well as numerous features of social organization" (Farnell 1999: 148).
>
> The anatomical, kinesiological, biological, and physiological sciences are not concerned with the syntactical or grammatical elements of dance idioms, nor are they concerned with the structured spaces in which dancers, priests, T'ai Chi masters, football players, Plains Indian sign talkers, and many others move. The transformational aspects of dancing as a symbolic mode of expression and communication are irrelevant in these kinds of scientific explanatory paradigms. And why not? We should probably think it absurd to imagine explaining a sculptor's or a pianist's artistry in terms of an anatomically functional analysis of his or her shoulders, arms, and hands, but we evidently do not think it is absurd with regard to dancers, for countless students have had to undertake studies of this kind in order to legitimate degrees in dance education, possibly because the woman who, in 1918, established the first dance program in the United States, as a part of the Department of Physical Education at the University of Wisconsin–Madison, was a biologist who had been deeply inspired by Isadora Duncan: Margaret H'Doubler (1962). I raise the matter here simply to underscore a point made some time ago: to have a database does not mean that one has a theoretical structure with which to approach it. (Williams 2004: 58–59)

So, what about the matter of kinesthetic awareness? Sweigard's work, I think, resulted in a kind of kinesthetic awareness, but as I have attempted to illustrate, she taught her work as a scientist, which meant that her teaching of awareness was *obliquely, not directly,* carried out. While her students certainly developed more acute awareness of their bodies and their movements, such awareness was treated as one of the by-products of learning ideokinesis—it was not a goal in itself. She would, I think, have said that studying ideokinesis produces (or opens the doors to) greater kinesthetic awareness, but that kind of awareness was not an end in itself with reference to her work.

It seems appropriate to say that because of the kind of training I did with Dr. Sweigard, I believe I became a more integrated person. The training I did with her helped me to mature—to grow up in important ways—but this kind of thing must also be understood as a by-product of the work we did together. She did not directly teach such things. While Dr. Sweigard welcomed reports of progress that included reference to the by-products I mention, she had strong aversions about over-emphasizing them, letting them slip out of their plainly subordinate role with reference to ideokinesis. In 2007, people tend to forget the massive resistance to Sweigard's and Todd's ideas during their lifetimes. Sweigard's methods were not welcomed in most universities, nor among physical educators back then. She was well aware of the fact that

> The ideokinetic approach to better skeletal balance through the use of imagined movement is a radical departure from the long established technique of relying on the volitional efforts of the individual to 'put' or 'hold' the parts of his [or her] body in better alignment. (Sweigard 1974: 213, cited in Williams 1981: 289)

Ideokinetic concepts are not parallel with, but counter to, many common notions of "good carriage":

> The old technique prescribes the practice of exercises to strengthen weak muscles which are often considered to be the cause of poor posture, and it accounts for the many admonitions concerning posture which have come into common use. Parents and teachers exhort children to stand tall, stretch up, put the shoulders back, hold the head up, and the chin in, tighten the "stomach" muscles, flatten the low back against the floor if supine lying, or against the wall if standing, and finally (the most reprehensible of all admonitions), to tuck the pelvis under. These exhortations are the province of the drill sergeant and the exercise master, but not of the educator. (Sweigard 1974: 223, cited in Williams 1981: 289)

They are also the province of the dance teacher, insofar as an idiom of dancing involves 'putting' or 'holding' the body in various positions. Often, a dance form will require a different beginning position of the axis of gravity (see Chapter 2, pp. 29–32, regarding mime and Kathak). Dancing is, after all, a social and symbolic activity that is learned, just as ordinary movements are learned as part of an expected code of demeanor, even when an individual (or group) may move in ways that rebel against commonly accepted codes of behavior.

I cannot overstress the fact that, to Sweigard, everything in ideokinesis began (and, to some extent, ended) in an individual's habitual postural pattern. In her essay entitled "The Dancer's Posture" (see Appendix to Chapter 3, page 48), she says,

> Posture is defined as the alignment of the body as it occurs habitually in the standing position with the weight distributed as evenly as possible on both feet.

> This definition recognizes a persistent and consistent position of the various parts of the body relative to each other and to a central vertical axis; also, that there can be but one posture or posture pattern for each individual.
>
> *Ideal Posture and Its Primary Values.* The unattainable "perfect posture" shows the skeletal framework in perfect alignment, in strict agreement with all principles of mechanical balance. Approaching the ideal promoted those attributes so important to the performing artist: (1) a slenderized figure; (2) optimum flexibility of all joints; and (3) minimum expenditure of energy, both in the maintenance of the body in the upright position and in the performance of movement. (Sweigard 1961: 39)

Notice the phrase "the *unattainable* perfect posture."

In a later essay written in 1964, published in 1965, she says "the dancer with the best posture will progress faster, reach a higher degree of attainment, and last longer in his profession" (Sweigard 1965: 1). Many dancers did (and do) not want to think that excellence in their dancing depends upon better postural alignment in everyday life. Moreover, when they found out that reeducating their postural alignment meant doing a lot of "brainwork," *mainly while not moving,* they were further disenchanted, often leaving Sweigard's classes to seek teachers who used traditional approaches. In the next chapter, these matters are explored more deeply, but here, I want to return to the idea of Sweigard's work itself in contrast to some of the by-products I've mentioned.

First, readers should try to understand that I believe many books about ideokinesis (and the so-called mind–body connection in general) to be books about possible by-products of the work, rather than ideokinetic teaching itself. For example, I don't think Sweigard cared much for Todd's book *The Hidden You.* She thought the book was a liability, as I do. To me, *The Hidden You* is an embarrassment, as is another book, called *This Magic Body,* by Jennette Lee (1946):

> *This Magic Body* is a record of my search for a technique to restore bodies to their natural balance and poise. It is training that engages both mind and body in a single act. It is not a form of mental healing; it is not physical training. It is both. It is a training of mind and body together, each having its allotted part, each equally important. (1946: x)

I have no idea who Jeannette Lee was, nor do I know anything about her educational background. She outlines a technique of training that includes the kind of imagery with which people who are familiar with Sweigard's ideokinesis are acquainted, and some of it would be judged "good" imagery by those who have had sufficient training in Sweigard's methods; however, the book is written defensively, from the standpoint of someone who was aware of the split between mind and body, and some of the stone walls of scientific theory and other traditions that insisted upon separating them.

Lee seems to write simply to reach as many people as she can, and her book may have reached many people. I don't know how the book was received during the forties, when it was published, but it is a good example of the dilemma in which people find themselves when they discover kinesthesia and the mind–body connection in their lives. However, in trying to write about their experience, something "other" tends to emerge. I've chosen a random example from Lee's book:

> The mind-body works by the universal process of life everywhere. A sound mind in a sound body means a vivid personality, a constant interplay of the forces that gave it birth, and still shape [*sic*] its work—the body unhampered by fear or worry of the mind, the mind trued to vital currents that flood up through it, filling it with gusto.
>
> And it would seem that, as in the body, power for a given act is generated through the contracting and relaxing of muscle-pulls that engage in the act, so in the process of thought, power may be generated by the interplay of mind and body in the thought-instrument—the brain focusing to a conception, the body relaxing to permit it to take vital shape. . . . The truth that I tried to make clear in the preceding section is not a theory, an idea that I made up to conform, in the world of literature, with the universal law of life. It is a truth that I feel I have a right at least to try to express since it has come to me through vital experience. (Lee 1946: 12–13)

I have problems with writing like this because I'm not sure I know what the author is talking about, or that I know what she means by such items as "the universalizing process of life everywhere." Others may not have these problems. Maybe they understand what she means. I don't, and with that confession of ignorance, I will return to Todd: I did not (and do not) expect to find any "me" that might be "hidden" in a book written by someone else, and I didn't find it in Todd's book *The Hidden You: What You Are and What to Do about It* (1976[1953]). No doubt some readers will disagree, but I found the title of the book off-putting to begin with, and its contents were equally unsatisfying.

Sweigard's book and her teaching methods were based on research and the years of work she did to explain Todd's fundamental idea—that mental images influence muscular action. Sweigard proved Todd's idea to be scientifically accurate. Her work took ideokinesis out of the realms of quasi-mystical flights of fancy, personality development, pseudo-psychologizing, and such.

The women in the classes I taught in Green Bay became kinesthetically educated because everything in those classes focused on a realm of sensory awareness they had not consciously encountered in their adult lives. Working with them, I proceeded from a fundamental Sweigardian idea: that of developing human movement *potential,* but I did no "table work" with them, nor did I undertake to correct individual postural patterns. Although some of their postural patterns improved—not as significantly as they would have done had they worked with Sweigard herself—there was improvement nonetheless. Above all, I didn't use ideokinetic concepts to affect how these women lived their lives, nor did I try to reveal a "self" hidden to them before they encountered the kind of movement education they acquired in Green Bay. For those who are interested in finding out what Sweigard taught along these lines, see "Good Mechanics in Everyday Movement" (1974: 261–73).

Having said this, I would not want to leave readers with the impression that Sweigard had no regard for dancers at all, because she did (see Appendix to Chapter 3). She would not have taught at the Juilliard School unless she had. Notice, however, that she emphasizes the connection between dancers and their ordinary *habitual posture,* thus providing us with evidence that her fundamental beliefs regarding the relationship between dancing and the underlying individual patterns of alignment in everyday life were her main message and the focus of a lifetime's interest. No more needs to be said about that here, as the reader can see what she wrote for him- or herself (see the appendix to chapter 3).

In the next chapter, we will discuss various applications of ideokinesis to professional dancing and dancers.

Appendix to Chapter 3
The Dancer's Posture

LULU E. SWEIGARD,
WITH INTRODUCTION BY
MARTHA HILL

During a distinguished career in research, teacher training, and private teaching, Dr. Lulu E. Sweigard came to the Juilliard School of Music, Dance Department in 1956, in the sixth year of that pioneer project in the professional training of young dancers—a pilot project by virtue of its introduction of serious study of the art of the dance in its various aspects into one of the world's great schools of music and by virtue of its aim. As Juilliard president William Schuman has written, "The musician [dancer] must not be content with technical proficiency alone. He must be equipped to contribute, through his profession, to the development of music [dance] as a constructive force in contemporary life."

To anyone who is familiar with the historical development of research and practice in the field of posture and efficiency in movement, Dr. Sweigard's work has always been in the vanguard of the mid-twentieth-century revolution in this field. Soundly based in the sciences of anatomy and body mechanics as well as neurology and physiology and drawing upon developments in psychology, her principles provide a common grammar for dancers and teachers of dance:

> "Movement resides in the thinking, not muscle action."
> "There is no voluntary control over muscle action."
> "Thinking and emotions influence muscle action."
> "Movement imagined in the body results in change in established patterns of neuro-muscular co-ordination."

The proof of the principles lies in the results achieved. Through Dr. Sweigard's meetings with the entire department, her laboratory sessions with the students,

---

Source: *Impulse: The Annual of Contemporary Dance* (1961), pp. 38–43.

her teaching of Anatomy for Dancers and her collaboration with dance faculty members, change in dance instruction and in accomplishment in movement go on continually. The results: more understanding of the instrument of the dance, the human body, and the medium, movement; realistic self-evaluation by the student of his own potentials and limitations; greater freedom and skill in movement; increasingly fewer injuries; faster recovery from strains.

The values of Dr. Sweigard's contributions to the Juilliard program are inestimable for the present and the future. These values are apparent in the increased understanding and efficiency of those working with her each year. Through her Juilliard alumni should come future advances not only in the craft of the dance, but, because of their wider perspective, gains in the art itself.

Martha Hill, director, Dance Department,
Juilliard School of Music

The fundamental efficiency of the normal human body as a mechanism for movement can be judged primarily by its posture or posture pattern. Posture is herein defined as the alignment of the body as it occurs habitually in the standing position with the weight distributed as evenly as possible on both feet. This definition recognizes a persistent and consistent position of the various parts of the body relative to each other and to a central vertical axis; also, that there can be but one posture or posture pattern for each individual.

Ideal Posture and Its Primary Values. The unattainable "perfect posture" shows the skeletal framework in perfect alignment, in strict agreement with all principles of mechanical balance. Approaching the ideal promotes those attributes so important to the performing artist: (1) a slenderized figure; (2) optimum flexibility of all joints; and (3) minimum expenditure of energy, both in the maintenance of the body in the upright position and in the performance of movement.

Deviations from Ideal Body Alignment. The deviations from good alignment in the central portion of the body, including the pelvis, lumbar spine, and upper thighs, are statistically more reliable (.91 to .98) than those in any other area (.73 to .87) when measured in relation to a vertical axis established in line with the center-top of the sacrum where weight divides to pass to the legs. Furthermore, these central deviations occur together consistently as a pattern of deviations in all individuals. And this pattern tends to persist throughout life, with a strong inclination to become worse with increasing age. This information was the finding of a study of bilateral skeletal alignment in the standing position as revealed in the x-rays of 500 subjects, divided into four age-sex groups.* It is logical that there should be a high degree of reliability of the deviations and their association in a pattern in the central part of the body since the center of gravity is located in this area and since the bones and muscles are of such large size.

---

*See Sweigard, Lulu E., "Bilateral Asymmetry in the Alignment of the Framework of the Human Body." Unpublished PhD thesis, New York University, 1939.

The Central Pattern of Poor Body Alignment. The pattern of deviation in the central part of the body is of interest to the dancer since it offers an explanation for the difference often experienced in ease and degree of movement of the legs. To describe the central pattern briefly, if the *right* hip (*ilium*) is higher than the left, the right femoral head and its socket on the pelvis are farther than the left from the vertical axis, mentioned earlier, thus giving greater lateral prominence to the right side of the pelvis, also, the pelvis is rotated to the right and presents a slightly greater antero-posterior tilt on the right side. This combination of pelvic rotation in the horizontal and sagittal planes is sometimes revealed by a slightly greater prominence of the right buttock; however, adjustment to it at the knee, ankle, and foot on the same side may be somewhat different in each individual. Finally, to complete the pattern, the lumbar spine deviates to the opposite side—to the left in case of higher right hip—usually in the form of a lateral curve of varying degree from person to person. If the *left* hip is higher, all the accompanying deviations are the reverse of those stated for the higher right hip.

Influence of the Central Posture Pattern on Leg Movement. The central posture pattern affects the relative freedom and range of movement of the dancer's legs. To check the difference in performance of the legs as claimed here, the dancer must first determine which hip is higher and more prominent laterally. Although an x-ray might give this information with greater accuracy, it can usually be determined by visible difference in the lateral contour of the pelvic area. If the *left* hip is higher and more prominent laterally, and the lumbar spine deviates to the *right*, the left leg will have greater range of movement than the right, its range being increased by further bending of the right lateral lumbar curve. The right leg, on the other hand, will tend to have its range of movement restricted by the resistance of the right lumbar curve to either straightening or bending in the opposite direction as needed to increase movement range of the right leg. Repeated and strenuous physical effort to increase its movement range will, more often than not, add muscular strain and will tend to build as a part of the posture pattern, another lateral curve in the opposite direction, usually above the existing one, but sometimes below it.

From the standpoint of mobility of the thigh joints themselves, when the left hip is higher, there will be greater mobility in the right thigh joint, also greater stability in single right-leg support with less lateral projection of the pelvis to the side. The dancer will tend to use this leg for support if he had a choice, and when he choreographs a dance he will tend to use the asymmetrical positions and patterns of movement which his own posture pattern favors.

To sum up the difference in the functional ability of the legs as it tends to occur in most dancers, the leg on the side of the higher hip will tend to have greater range of movement, but less stability in one-leg support; the opposite leg will have less range of movement, but greater stability in one-leg support.

Tests of Flexibility. Among many possible tests of body asymmetry and flexibility of joints there are two which are often used in our posture laboratory classes to familiarize the dancer with some of his problems of body alignment and the accompanying restriction of free movement. Other things being equal, the dancer of slender build will invariably have an advantage of the one who is more heavily set.

The first position tests the flexibility of the main weight-supporting joints of the body, with all due respect for the principle of two-joint muscle action. The joints involved are those of the entire spine, the pelvis, and the legs. To assume the position, the dancer kneels with the knees in vertical alignment with the thigh joints, shoulders and arms hanging easily, and feet extended. He then sits back on his heels and, while maintaining the contact of his pelvis with his heels, bends sharply at the thigh joints (flexion of the trunk on the thighs) to allow his forehead to rest on the floor *beyond* the knees. The spine should then present a long backward curve, slight but even in degree throughout its entire length, the shoulders should hang forward toward the floor and the arms rest on the floor at the sides of the trunk. Areas of restriction of bending and those of an undue amount of bending in the spine are readily detected in this position. If the dancer cannot assume the "fold-up" position, meeting all its requirements with ease, there can be but one reason: habitually tight and sometimes bulky muscles which will not lengthen and soften to permit the full range of bending allowed by the nature of the joints involved. It should be recognized here as in all positions and activity that movement in the joints is restricted first by muscles, secondly by ligaments, and finally by the structure of the joints themselves.

Trying to force the body into this, or any other, position does not decrease the tightness of the muscles restricting the movement; it increases it. This is due to the operation of one of the principles of muscle action, the stretch reflex. When a muscle experiences pull or tensile stress by position, movement, or an outside force, its contractile fibers shorten to resist the pull. Hence, repeated, forcible stretching of muscle increases its contraction and tightness, which becomes evident in some areas in bulging contour, in others as hardness of muscles resembling that of bone. The latter is especially noticeable in the low back (lumbo-sacral) area. Even so, dancers know they have increased range of movement by forcible stretching of muscles, the hamstrings, for example. Such stretching, however, has not occurred in contractile fibers of the muscle, but in the connective or fascial tissue which binds contractile fibers together and attaches them to bone. To retain the length secured by forcible stretching requires constant practice of the same procedure, but in this, enthusiasm should be tempered by judgment to avoid tearing of soft tissues.

The second position used in the laboratory further tests the flexibility of the thigh joints and those of the spine, especially in its lumbar portion. Few dancers can meet the requirements of this position even though the movement needed in the joints is well within the range allowed by their structure. To assume the position, sit on the floor with the legs to one side; this requires inward rotation in one thigh joint but outward rotation in the other. If the legs are to the right, the sole of the left foot should just touch the right thigh along its lower part; the right knee should be bent from the pelvis. Ideally, both sides of the pelvis at the tuberosities of the ischia and both knees should rest on the floor, while the trunk balances upright easily and without lateral bending.

This position is influenced by the central posture pattern as follows. When the right hip is higher and more prominent laterally, the position is assumed with greater ease and conformity to its requirements when the legs are to the right. When the left hip is higher, the reverse tends to be true. In this position, however, the firmness of the lateral deviation in the lumbar spine and the relative degree of outward rotation of the thighs (which showed no consistent relationship to the central pattern of body alignment in the x-ray study) can at times modify the influence as stated. Therefore, it is possible for a few individuals to experience less difficulty when the legs are placed to the side opposite the higher hip, and for some to experience no preference regardless of the direction in which the legs are placed.

In trying to assume the sitting position with the legs to one side it will be found that all attempts to force one side of the pelvis to the floor will result in raising the opposite knee higher from the floor, and vice versa. This illustrates the principle that forcing one part of the body into a desired position results in a corresponding deviation of some other part or parts of the structure in accordance with the pattern of muscular limitation of free movement in joints. The principle holds true in all patterns of movement although its effects are not so readily detected in standing, especially on one foot. It accounts for many of the faults of position and movement which plague the dance teacher. One very good example can be noted in the turn-out position of the feet in which the movement should occur at the thigh joints. Forcing the turn-out of the feet when there is restriction of movement at the thigh joints results, more often than not, in deviation from good alignment at the joints of the knee, ankle, and foot. Thus repeated effort tends to promote knock knees and inward rolling of the ankle and foot (pronation of the foot).

The Posture Laboratory. The supporting facts for the many principles which guide the procedure used in our posture laboratory are found mainly in Anatomy, Neurology, Physiology, Psychology, and Mechanics. Only those facts which are responsible for our omission of all traditional posture and conditioning exercises will be mentioned here.

1. *Gray's Anatomy** states that movement resides in the thinking, not muscle action. This presents a warning to both the posture and the dance teacher: teaching should be done in terms of movement. It is futile, and also detrimental to neuro-muscular coordination, to tell the individual what to do with his muscles, even though the teacher may have a phenomenal amount of knowledge of them and their intricate and complicated work. The more one knows about neuro-muscular physiology, the easier it is to realize that the fast-changing patterns of muscle action involved in movement patterns and in maintenance of body equilibrium will be forever difficult to assess.

2. *There is no Voluntary Control over Muscle Action.* This principle re-enforces Gray's statement. A familiar illustration of the function of both principles is

*Gray, Henry. 1930. *Anatomy of the Human Body.* [22nd Edition]. Revised and reedited by Warren H. Lewis. Philadelphia: Lea and Febiger.

seen in the way a baby learns movement through a trial-and-error process. Before the age of five years he has established the basic neuro-muscular coordination needed to maintain his upright position and to produce all the movements allowed in his joints, as well as many of their combinations, in daily activity. If he later studies the dance he will no doubt increase his range of movement in various joints and learn new patterns of movement, but not new movement.

3. *Thinking and Emotions Influence Muscle Action.* This is a maxim which is quite generally accepted today, but in a somewhat negative way. If thinking can have a detrimental influence on muscle action, it can likewise have a constructive influence, if it can be directed. Inherent in this statement is a very significant warning to the dancer himself: repeated and strenuous physical practice of a position or movement pattern is not enough; it must be supported by constructive thinking if there is to be adequate improvement in performance. To be most effective, however, thinking must be based on facts. For the dance teacher, the more he can challenge the dancer's thinking with pictorial and accurate description, along with demonstration of positions and movement patterns, the better the response of the dancer.

   In the posture laboratory there is time and opportunity, as presented in no dance class, to put this principle to work. The way is through education. In our laboratory sessions, facts from various fields of knowledge pertinent to the lesson of the day are presented—a continuous story from lesson to lesson. Thus the student accumulates factual information about his body as a mechanism for movement. Information that seems most important in his dancing is the following: knowledge of good body alignment and its values; knowledge of his own body alignment and its restrictions on movement; the exact location of the thigh joints; the movements allowed in the various joints; the limitation of extension of the thigh by the ilio-femoral or Y ligament; where movement occurs in various fundamental positions and patterns of movement of the dance; and lastly, principles of muscle action. There is no mention of muscles themselves except, at time of their grouping, as the hamstrings or abdominals, or to clear up some misconception about them. Finally, it is important that the dancer know and practice good mechanics in his everyday movement, for this can contribute to better dancing. Needless to say, a skeleton is a constant aid in all laboratory teaching.

4. *Movement Imagined in the Body Results in Change in Established Patterns of Neuro-Muscular Coordination.* There is no better proof of the validity of this principle than the student's experience in the posture laboratory. *Where* imagined movement should be located and what its *direction* should be is based on the study of approximately 300 subjects. In this study [from unpublished data], a series of repeated measurements of body alignment were

> made before and after a period of teaching in which imagined movement was substituted for voluntary movement. The changes which occurred in the relationship of parts of the body to each other, and to vertical and horizontal lines, indicated improved conformity of body alignment to principles of mechanical balance. Those changes which occurred in the same direction with the same location in all subjects were considered to indicate the location and direction of the imagined movement which would promote change needed in basic neuro-muscular patterns of coordination to produce a better body alignment. Imagined movement is first tried in the back-lying, knee-bent position (the constructive rest position—CRP), later in many other rest positions and in standing and walking. Some few imagined movements which are most far-reaching in their effects on the body are set up for use in the dance class during brief respites from activity.

To give a more complete picture of our laboratory work, it should be pointed out here that voluntary movement of small degree is used as soon as the student has been somewhat successful in changing muscle coordination by imagined movement only. Voluntary movement relates to the shoulders, the thighs, the feet, and the head, separately, and in various combinations. All such movement, however, is dominated by mental activity. Once the technique of the movement is learned, the student is free to concentrate on imaged action to promote and maintain better alignment and integration of the trunk as voluntary movement proceeds. Special attention is given also to breathing, alone, and as it can be used to favor integration of the trunk during movement.

Since all principles of neuro-muscular function suggest increase in existing problems of muscle strain when there is continuous, rigorous use of the body, and since the dancing profession, as perhaps no other, engages in many patterns of movement which defy both principles of neuro-muscular function and principles of good mechanics, it appears reasonable that work in the posture laboratory should accompany the training of the dancer. It has been shown that such work can improve the efficiency of the bodily mechanism, even as it is subjected to strong use. Ultimately, the responsibility for improvement rests with the dancer. He must be willing to engage frequently in concentrated, constructive thinking, both to nullify the effects of his strenuous physical work and to increases the efficiency of basic neuro-muscular habits. Improvement comes slowly, but surely, and its value multiplies with advancing years in the profession.

Chapter 4
Doctors, Dancing, and Ideokinesis

Before making specific distinctions between teaching dancing to professionals (or aspiring professionals) and the kind of classes I taught in Green Bay, Wisconsin, I will make a further point about Sweigard's book and her system of teaching as I conceive it. I alluded to various responses to her book (and, by extension, to ideokinetic practice) in a review written in 1979, published in 1981:

> If one thinks of dancers reading the book, one wants to say, do not reject it or the method because you do not understand the language or do not think of your bodies in this way. The language in which the book is written is, after all, a respectable language although it may seem far removed from concepts of beauty and other more artistic or aesthetic notions about the body which you may have. (Williams 1981; 292)

Part of the reason that dancers (and others) don't read Sweigard's book can be attributed to the language in which it is written, which is challenging for people who haven't been trained in any of the "body sciences." She makes this point herself in Chapter 2 of her book, saying:

> *In studying any subject, the student must first learn the language used in the field.* Here the student must become familiar with the names of various parts of the body, the skeletal structure in particular, and must be able to describe and locate these parts. He must know the kinds of movement that occur in the joints of the skeleton and the names applied to these different movements. Terms such as *hips, hip joints, backbone,* and *stomach muscles* usually are avoided in this text because they often promote inaccurate concepts or vagueness concerning their location in the body. For example, *stomach muscles* have nothing whatever to do with voluntary movements: the stomach is an internal organ. *Hip* and *hip joint* are anatomical terms, but most people, even some who have studied anatomy, cannot locate them accurately. (1974: 8—italics added)

In my understanding (when teaching years ago and now), Sweigard recognized the difference between the "anatomical" (medical) lexicon of the body in contrast to the "social lexicon" of the body (see Appendix to Chapter 1, page 18), explaining why she didn't use the (English) social lexicon in her book—it was inaccurate in terms of movement.

When I was teaching, many dancers (and students of dancing) didn't want to learn the scientific terminology of the body, nor did they want to learn the movements of which their joints were capable. I am told that now this is not the case, although I remain skeptical regarding the extent to which the entire American dance world has changed. I believe it has changed in small enclaves in metropolitan centers like New York and a few universities, but I doubt that a general linguistic revolution of this kind has taken place throughout the country. Be that as it may, we will turn now to the viewpoint of part of the scientific community toward ideokinesis as it existed when I was teaching. Such discussion naturally involves complementary attitudes among dancers and teachers of dancing.

Of particular concern then (and now) were professional medical attitudes toward movement education, for *any technique or system that bases itself on a notion of maximization of human movement potential* is likely to be looked on askance by doctors—at least, it was in the late fifties when I studied neuromuscular reeducation. This is unfortunate, for the sanction of reputable members of this profession is tantamount to credibility and the acceptance of ideokinesis by many people, but there is a difficulty. Why?

In general, medical doctors are trained to bring people from states of malfunction, dysfunction, and/or disease to normal states of health, which (for our purposes) may be defined as freedom from pain. A movement educator's or dance teacher's aims, problems, and objectives are not the same, because the latter are guided by notions of *optimum function* and the *maximization* of movement potential. They begin their work with normally healthy bodies. In its broadest definition, ideokinesis is about maximizing the human potential for movement. *It is not about malfunction.*

If traditional medical criteria of sickness and wellness are applied, then it can easily be seen that medical criteria of "normalcy" and "freedom from pain" provide no categorical slot for, or explanations of, *superlative* performance (Williams 1981: 292—italics added). But medical doctors deal with the human body, and so also do movement educators and dance teachers. Nevertheless, the doctor's concern with living bodies is unlike that of movement educators, because nearly all teaching of dance forms tends to begin where medical doctors leave off.

Sweigard said: "the dancing profession, as perhaps no other, engages in many patterns of movement which defy both principles of neuro-muscular function and principles of good mechanics." She was (and still is) right, and while it may be true that more awareness of dancers' problems exists in the medical profession now than fifty years ago, there is still a long way to go.

In general, doctors are not required to do anything more than bring an injured, diseased, or abnormal living body to a state of wellness, generally called normal or average: a state of being that is meant to include freedom from pain. It therefore bears repeating that movement educators (whatever they are teaching) *begin* from the medic's perception of *wellness* or normalcy. Medically approved and practiced therapies of various kinds simply don't enter into the picture; therefore, it is relevant to ask, "What are we talking about?"

It would be good to know, for example, what the objections were that prompted a New York medical association to bring a lawsuit against Mabel Todd in the fifties (see Matt 1993: 50), because the allegations made in the lawsuit would serve to define, in part at least, what the association's objections were. What aspect(s) of Todd's work did the association believe infringed upon medical practice? The allegations would provide insights into some of the issues under discussion.

Movement Pathology versus Movement Potential

In teaching contexts, I used (and would still use) the following diagram purely as a heuristic device when I attempted to explain issues created by the spectrum of possibilities between movement pathology and movement potential:

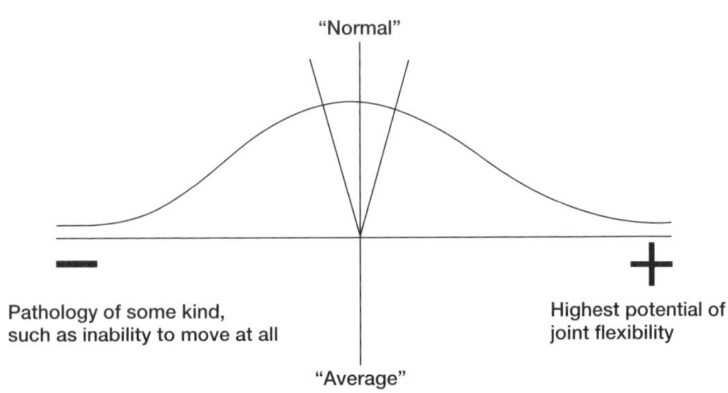

Drawing 2. Bell curve.

I apologize for the crudeness of the diagram (not a good illustration of a "Bell curve," among other things), but it provides enough of a picture for us to get on with. First, however, I must say that when I use the words *movement potential*, I think of the potential displayed by professional dancers, gymnasts, or ice-skaters. When Sweigard spoke of movement potential, she was thinking of ordinary, everyday movement.

Medical doctors and physiotherapists work from the minus side of an imaginary spectrum toward the centerline. Some of them doubtless go a small way into the plus side (indicated by the right-hand diagonal line around the field designated as normal), but I think it safe to say that they do not, in general, proceed further than this—if they get that far. On the other hand, movement educators (including dance teachers) of all kinds begin where medical professionals end.

It is true (as you will soon see) that some dancers *start* from a point somewhere on the minus side of the spectrum (the area marked by the left-hand diagonal), perhaps because they have bursitis or lower back pain. Perhaps they have suffered pulled or torn knee ligaments or cartilage, or they have broken a bone or something. After being treated by a doctor, they seek to work with a movement educator or a dance teacher who has sufficient anatomical and other relevant knowledge about their idiom of dancing, so that they can rehabilitate their problems in terms of the already increased movement skills they have acquired throughout their lifetimes.

Would-be professional dancers are people who are determined to *surpass* the normal range of movement required for everyday life. Those who have attained the status of professional dancers have achieved this level of maximum physical capability. They have pushed the normal limitations of their bodies to an extreme required by professional ballet dancing or some technique of modern concert dancing. Therefore, when asked, "How does neuromuscular reeducation or ideokinesis help such dancers?" many are surprised by my answer: Ideokinesis can help professional dancers *in limited ways,* but it is of less assistance to them (and to figure-skaters, gymnasts, yogis, acrobats, or contortionists) than it is to other groups of people.

Ideokinesis helped aspiring dancers to overcome (relatively minor) postural problems. Sweigard worked with many of these at Juilliard. To her students, the problems were seen as those which prevented them from attaining the ideals of performance ability they envisioned. Ideokinesis always was (and still is, on the whole) in an underlaborer's position with regard to professional dancing.

We can safely say that, in the dance context, ideokinesis has little or no value to most dancers in (and of) itself. That is, on the whole, *dancers do not undertake ideokinetic study unless they are hurt.* All that a seasoned, or an aspiring, dancer wants of ideokinesis is enough knowledge and rehabilitation to get on with his or her dancing. That aim already removes Sweigardian ideokinesis from its intended purposes to become something it was never intended to be. Instead of movement education, it is turned into a vague, (usually) temporary cure of some kind.

Successful professional dancers generally have no need of the kind of help ideokinesis offers with regard to the correction of postural aberrations such as severe scoliosis, pronated ankles, and such. Had they had such problems, they probably would not be dancing. There is little or no possibility for persons so afflicted to be (or to become) professional dancers in any case.

Conscious Tension Reduction

Perhaps the greatest thing that ideokinesis has to offer professional dancers lies in the area of *conscious tension reduction* in the CRP. Ideokinesis provides professionals and aspiring professionals with ways and means of controlling bodily tensions that, to my knowledge, is unsurpassed. Another benefit of exceeding value is that it assists them in *separating the moves they make while performing from their habitual moves in everyday life.*

Then, too, if a dancer has injured him- or herself dancing, ideokinesis can aid the healing process in important ways through sharper, more effective imagery pertaining to the injured part, but I did not recommend the study or practice of ideokinesis in cases other than these.

As examples of why I say this, I offer teaching experiences I had with three professional dancers:

1. Lucas Hoving was a well-known modern concert dancer who came to me unable to raise his arms above shoulder height owing to severe bursitis in both shoulder joints. He was about forty—possibly forty-five—years old. Unless his problem could be resolved, he was looking at an early end of his career. He had been to several medical doctors with his problem, who, without exception, prescribed pain relievers, although Lucas told me that one doctor did suggest that he quit dancing.

   I worked with Lucas every week for two years, telling him that there was no such thing as a quick fix. The major part of the work I did with him was on tension reduction. As he slowly became able consciously to control the tensions in his body, we began intensive work with centering images (the light beams) throughout his body, especially in the arms, neck, and shoulders. He learned what the actual bony structure of his body looked like, and he learned accurately to locate bodily parts. He learned the kinds of movement of which his joints were capable. He also learned how *not* to

> carry performance tensions in his body during the time he was offstage or in classes.
>
> He discovered through this process (among other things) that for most of his professional life, he had confused muscular tensions in his neck, shoulders, and upper rib cage with good expression while dancing. He found he had to separate his notions of expressive movement from muscular tensions in those areas of his body.
>
> Throughout the process, I insisted that he work *without mirrors* (standard pieces of equipment in most dance studios and rehearsal rooms). Lucas found this requirement extremely difficult, but when he realized that some of the tensions in his upper body had been accumulated through the years by watching his mirror image rather than paying attention to the movements and their internal (kinesthetic) effects themselves, he perceived the value of work *without* mirrors, and, after two years, he was able to perform more effectively than he had done prior to the problem with his shoulders. Moreover, to my knowledge, he never had such problems again.
>
> I also worked with Lucas on one of the roles he had to produce in performances: the Christ figure in José Limón's *The Traitor,* which was extremely difficult because it required a sustained continuation of an extreme forward-leaning axis of gravity throughout the entire performance. The artistic reason for this was that the Christ figure had to appear not only to be *above* everyone else on stage, but to *hover* over all the performers who danced with him in *Traitor.*
>
> The forward-leaning axis made different demands on his body and its musculature throughout every danced move he made as this character. However, given the control he gained over bodily tensions in the CRP (which he used before and after every performance of *Traitor*), and given the centering imagery for the movements themselves, he was able successfully to perform the character without further damage to his body, although it was by far the most strenuous role he performed.

Hoving's case is instructive because of a point made in Chapter 2 about mixing Sweigardian images into those required in a discipline of dancing and/or mime. If I had tried to teach Lucas to *translate* ideokinetic images for ordinary walking *into* the images necessary for the posture, moves, and walking he had to do as the Christ figure in performance (or, for that matter, any performance), it would have been wrong to do so. Instead, I taught him to reduce the tensions in his body such that he didn't carry them from the stage into his ordinary life, and I taught him different imagery to use while he rehearsed and executed the performances of that specific character (see Chapter 5, section on 'Mental practice,' for the kinds of things we did).

2. Liane Plane was a female ballet dancer who could scarcely bend over to pick up her *pointe* shoes—far less dance in them—when she came to me. She had finally injured her knees so badly during a performance of *Harvest According* that she had to find some way to repair the injuries (pulled muscles, tendons, and ligaments in both knees, the right being worse than the left) or quit dancing entirely.

   Liane and I worked together for roughly two years. As in Hoving's case, I insisted that she work without mirrors. For a bit more than a year, the emphasis was mainly on tension reduction and the establishment of Sweigard's lines of direction and action in the CRP, and then we progressed to side-lying and sitting positions. When we were able to achieve ordinary walking in terms of ideokinetic images, we could address her fundamental problem, which consisted of nearly everything she did in ballet technique classes, hence onstage. Her early dance training was seriously flawed, mainly because it proceeded from the notion that practicing exercises to strengthen weak muscles was the major aim of a ballet technique class. She had also been taught ridiculous ideas about "turn-out," trying to force her legs into unnatural positions.

   Liane's problem with her knees (and subsequent referred pain in her back) was caused by poor ballet training and carelessly taught technique classes during most of her life and as a professional. In other words, the structure of dance technique that she had built up over the years finally took its toll on her body—as it always does. She was, however, a strong dancer, with brilliant movement attack, good jumps, and fast turns. In Noverre's terminology (1930[1760]: 35–36), she was a typical *arqué*-legged dancer, so she survived in spite of it all until her mid- to late twenties.

   Although it was difficult for her emotionally, she literally had to relearn basic ballet technique. We went through the whole syllabus of moves, starting with *pliés,* completing everything pertaining to *barre* work and *port-de-bras,* and then we worked on center floor exercises. It took a long time for her to develop patience and to overcome the accumulated anger she felt toward past teachers who didn't really know what they were doing. Many times, she would pound the *barre* with her fists in frustration, saying, "But *why* wasn't I taught that in the beginning?"

   Her fundamental problem with ballet dancing was with the turned-out leg positions—not an uncommon problem among professionals:

   > I have a vivid memory of one of the teachers stopping me in the middle of class to demand that I turn out my feet. There was no regard for the knees or hips, which in my case were distorted to the breaking point. The teacher refused to continue the class until I complied with her wishes.

> . . . During my third year of training, I was introduced to toe shoes, the standard footwear for all ballerinas since the mid-nineteenth century. . . . Misguided and caught between excessive demands for turn-out and *pointe,* my feet had already begun to deform. At the age of eleven, I came down with a severe case of bunions. Many of the teachers had the same malady, caused from years of strain placed on the foot. It was said that Balanchine cherished the aberration of line induced by bunions, that they contributed to the impression of winged feet. (Kirkland 1986: 34–35)

Liane and I worked with the imagery described in Chapter 1, and we worked with images of the pelvis, the femur, the tibia and fibula, and the tarsal bones of the foot. She also had to develop different images of basic balletic movements. Liane had to get rid of the notion, cultivated unconsciously through the years, that ballet dancing was mainly composed of muscle work and force. The anatomy lesson I taught her was simple:

> *BONES MOVE THROUGH SPACE; LIGAMENTS HOLD BONES TOGETHER; TENDONS ATTACH MUSCLES TO BONES, AND THE MOVES YOU WANT TO MAKE ARE CONTROLLED BY IMAGES THAT SEND MESSAGES TO MUSCLE GROUPS, NOT SINGLE MUSCLES.*

At first, when we were finally able to start her balletic retraining, she would ask, "Now, what muscle do I move?" When I told her that *there was no single muscle* that finalized any of the moves she wanted to make, she had problems giving up her received ideas, but she did, and we succeeded. After Plane and I finished our work, she got a job in a Broadway musical understudying Chita Rivera in *West Side Story*—a real victory for her, because the kind of movement she had to do (commonly known as Broadway jazz) is especially hard on knees; however, as far as I know, she never had any more serious problems.

Liane Plane's case is instructive in four ways: it points to 1. the differences between Sweigard's imagery and the imagery necessary in a ballet technique class; 2. the understanding that ideokinetic knowledge demands that students give up previous, often misguided, ideas about the body and cultivate new perceptions about how their bodies function; 3. the fact that an entire dance technique can be taught in such a way that it makes injury inevitable (see Kirkland quote above), even if the habitual postural pattern isn't all that bad; and 4. the fact that there are no quick fixes in ideokinetic terms. The kind of rehabilitation that Plane needed took more than two years to achieve. I often said to her, "Look! It took roughly fifteen years for you to get into this mess. A couple of years to get out of the mess is short by comparison."

Liane's experience also points to the fact that ballet technique taught from a standpoint of nothing but standard, often-heard "postural admonitions," such as "turn out your feet," "tuck your pelvis under," and such simply don't work.

3. José Limón was a well-known, distinguished star of the modern concert dance world. He came to me when he was nearing fifty, because he was troubled with chronic lower back pain, which seemed only to get worse as he got older. The cause of the lower back pain was a constantly hyperextended chest. When I pointed this out to him, he agreed that I was probably right, then asked what, if anything, I could do about his problem.

   I told him that I could teach him things that would alleviate his problem, emphasizing that in so doing, I would teach him to change his posture entirely during the time he was offstage. I pointed out that he would have (literally) to learn to walk again, reserving the hyperextended chest for use in the roles he performed onstage *only.*

   I told José that he would have to learn tension-reduction techniques, and different ways of sitting, standing, and walking from his habitual ways of performing these acts, all of which were based on how he danced. Limón's posture was the same in ordinary life as it was when he danced onstage. Since that was the root of his problem, the only way he could get lasting relief from the pain was to break the postural habit during the times he wasn't dancing.

   I did only five lessons with José because he realized, after that many sessions, that my work could indeed effect the changes in his postural habits that I said it could; however, he then had to face up to a bigger (and to him, insurmountable) problem: the hyperextended chest that, to him, was an important embodiment of his masculinity. He told me he couldn't imagine himself with his chest in any other position in relation to the rest of his body. He said that the hyperextension was a hallmark of his style and that he was afraid of losing it if he learned to function differently in ordinary life. He was unable to make the adjustments that Lucas Hoving had made by separating that which he performed onstage from his "performance" in ordinary life. He felt that he would somehow lose an important element of his identity.

   I told him I understood what he was saying and that the decision of whether or not to continue studying ideokinesis was entirely up to him. I said I wasn't prepared to handle the psychological difficulties to which he alluded regarding the identity problem, although I was fully aware that there would be deep psychological aspects to making the change in his overall posture that I told him he had to make. It was up to him, not me, to decide whether he could cope with his problem if he aimed to get rid of the back pain.

   He decided the risk was too great. After all, he was the best judge of his powers and capacity to address the matter. I made no attempts to persuade him

> otherwise. We parted friends, but I didn't teach him again. He sent some of his own students to me, telling them that I was a "back-to-the-womb teacher," which I took both to be a compliment and his interpretation of the penetrating nature of Sweigard's ideokinetic *principles* that were utilized in my method of teaching dancing.

José Limón's case is instructive because it points to the highly individual nature of ideokinetically based work. *It is not a universal panacea,* nor should it be understood to work the same way, in all cases, at all times, with everyone. Furthermore, José's case illuminates the fact that postural aberrations (the hyperextended chest) can originate not from anatomical disorders such as, say, pronation of the feet, or mild scoliosis, but from the images an individual has of his or her social and/or stage persona.

Limón had no doubt that ideokinesis would work in his case, but he felt he couldn't cope with the changes it would effect in his life as a whole. This in no way blames or detracts from him for quitting the study, because every individual differs in terms of how much change they can absorb in their lives at any given time.

In strong contrast, Lucas Hoving managed to make the separation between moves he made in his ordinary life and the moves he made onstage. Liane Plane, on the other hand, had to relearn the dance technique she relied upon up to the time of her injuries. She had to relearn her art in new and different ways.

The changes that working from ideokinetic principles brings about in an individual person's life are often profound. The ways in which students see their selves have to change. I went through many changes as a result of working with Dr. Sweigard, but I don't believe that the same work (or combination of images) would effect the same changes in other people. It will suffice to say that long-term, habitual patterns of movement and posture change as a direct result of studying movement informed by ideokinetic principles (see Appendix to Chapter 4, page 69, for Sweigard's analysis of the relation between posture and dancing). The concept of self extends to every aspect of life. People sometimes say, "My life has changed," but in reality, their concept of *self* has changed, and that changes their lives. To conclude these comments, I turn again to my review of Sweigard's book:

> Dr. Sweigard was not engaged in the currently extremely profitable business [in the sixties, seventies, and early eighties] of being a 'guru.' She was not a quack doctor, a woolly metaphysician, a 'mystic' or a crank. She resolutely refused to comment, even to speculate, about the consequences of her work outside the boundaries of her certain knowledge as anatomist, kinesiologist,

and movement educator. As I was among her more obstreperous pupils, I taxed her for this.

The questions regarding interdisciplinary applications of ideokinesis (or possible attitudes towards it) were not answered then and it may be that they will not be answered now, but I would want to say the following things: Her book will only be unfairly criticized by philosophers because no philosophy of mind is made explicit in the book, although it will be obvious to most of them that ideokinesis *implies* a philosophy of mind. It would be of great value if a trained philosopher could take inspiration from it, learn the method and explicate for us just what kind of philosophy of mind ideokinesis entails. (Williams 1981: 291–92)

The Limitations of Classroom Techniques

Further to the point of teaching dancing to professionals is the relationship between ballet classroom techniques and the artistry required for performing a role onstage. Both classroom and stage dancing require imagery, but of a different kind. Many people, including many dancers themselves, believe that years of practicing classroom exercises will somehow, in the end, reveal the *meanings* of similar moves when they are utilized in performances onstage, but classroom techniques are not designed to do this.

Perhaps only highly trained, accomplished performers are able to understand Gelsey Kirkland's assertion that there was "a disparity between what I felt and the feeling that I was walking on a cloud or floating across the stage" (1986: 131). Understanding that different kinds of imagery are involved in the sensations a performer feels while onstage, and the image that he or she creates for the audience, is knowledge tantamount to fine performance, but it points to a highly sophisticated conception of performance in dancing—and a highly sophisticated conception of imagery in general. Kirkland nails the disparity between these types of imagery in unusually clear terms:

The real feeling was frequently in opposition to the quality embodied in the movement. Even a simple movement, like rising on *pointe,* demanded that I exert a force in the opposite direction, *pushing down against the floor,* resisting the inclination to *pull myself up* with legs and arms. Poetic inspiration, what seemingly lay behind the physical imagery of the dance, did not reveal the method of composition. (Kirkland 1986: 131—italics added)

Kirkland draws an important distinction *between the visual imagery meant to be received by an audience* for the characters dancing onstage and the dancer's imagery that creates those images for an audience. She rightly says that images such as "push the floor down" produce totally different *aesthetic* effects regarding the movements, which are "floating in a dream" or "walking on a cloud." However, if the dancer visualizes "floating in a dream," she will not convey that image to an audience. And Kirkland is right: a complex linguistic process of translation has to occur for a dancer adequately to produce the desired aesthetic or dramatic images in performance.

At the end of Chapter 5, I will undertake a more complete analysis of the relationship between these two types of imagery. Suffice to say here that exceptional performers use other kinds of imagery to create the aesthetic surface of a dance performance. What is unfortunate is that an audience often pays more attention to the sensuous surfaces of the dancers than to the aesthetic surface of the dance being performed. There are two factors involved, both of which Kirkland mentions:

1. "I never received an explanation of the stories behind these ballets," she says. "Each ballet had to be investigated moment by moment. . . . Each step had to be broken down into the basic physical principles that had to be applied in order to compose the movement. Then the steps had to be phrased, according to the same principles" (1986: 130).

2. "The technical key for which I was searching—which would link the dance to the music and to the drama—was known as 'legato.' This was the special dynamic which seemed to allow the Russian dancers to move through the steps with such fluid grace, to bind all the steps within the overall concept of the ballet. The Russians themselves, including Misha [Barishnikov], were unable to translate the verbal cues and corrections which they had received in their training into a comprehensive and intelligible idiom. They were often unable to communicate the knowledge that was stored in their bodies to American dancers, whose training was geared to a different aesthetic. 'Legato' was an alien concept to Americans who had been conditioned for speed, blurring rather than binding the steps" (Kirkland 1986: 131).

She points out that Barishnikov danced "following an elaborate set of rules that originated with the Russian classroom syllabus" (1986: 131), thereby emphasizing a point that escapes the majority of laypeople regarding a dance technique: the overall, *cumulative design* of the technique. The overall, cumulative design of a ballet technique consists of the composition and organization of the *whole* technique—a structure that is much more than an aggregate of its separate elements.

Summary

At the end of Chapter 3, I expressed the hope that readers would follow me from the moves required for baking biscuits in Sweigard's kitchen to the more complex movements required of professional dancers. During the process of getting to the complex movements required of professional dancers, I've asked readers to consider the interface between movement potential and movement pathology and how ideokinetic principles as I conceived of them entered into the lives of three professional dancers who worked with me because they were injured in one way or another by their dancing.

Thanks to Gelsey Kirkland's struggle with the imagery necessary for performing ballets onstage, I've tried to push the complications of the movements of professional ballet dancers to the limits required by their art. From baking biscuits in Sweigard's kitchen (requiring imagery), we advanced to movements required of a professional dancer onstage, also demanding imagery, but of (at least!) two, and probably more, different kinds.

The implied spectrum of movement is mind-boggling in terms of its complexity, range, balance, and strength. Yet, as I trust the discussion so far has made clear, significant aspects of the problems experienced by Lucas Hoving, Liane Plane, and José Limón (and they were, and are, not alone) lay in the fact that Hoving and Plane had in some sense lost touch with their everyday movement. Both had to be retaught everyday moves in relation to their performing abilities. Not only did they have to recognize and place importance on their everyday movement, they had to separate these moves (all the moves they made offstage) from the moves they performed in classes and onstage. Limón's problem was somewhat different: he simply moved as a dancer all the time, no matter whether he was performing or not.

We give lip service to the intricacy of the professional dancer's art but do not pay close enough attention to the real scope of the problems involved—or the fact that many of the problems are connected to the dancer's everyday posture. Similarly, we give lip service to something we call ideokinesis, but we don't seem to pay attention to the consequences (possible or actual) of teaching it, *why* we teach it, or *what* we are teaching that is labeled ideokinesis.

I would want to say that ideokinesis has in some sense become a system, because of Sweigard's book and the fact that she worked with dance students at Juilliard in a posture laboratory, hence there are many who believe ideokinesis will assist *all* dancers with their bodily problems no matter *what* those problems may be, *how* they may have arisen, or what the consequences of deep postural, hence psychological, changes will be. This is to say that I believe ideokinesis in 2007–8 has become falsely universalized, to the extent that the word is used in so many ways that it no longer has meaning that ties all the usages together. It is now what Wittgenstein would have called "a portmanteau term."

In the interests of clarity, we will examine a few uses of imagery and movement, contrasting them to Sweigard's method of teaching. Readers may well ask, "What *is* the relationship between imagery and dancing?" It is a significant question because it raises issues regarding the use of imagery while the body is *passive,* in contrast to using imagery while the body is *moving:*

> According to specialists in the area of skill acquisition (Suinn 1976), the benefits of mental imagery on dance learning or performance are maximized when the dancer remains perfectly still. Dance experts, however, beg to differ (Dowd 1981). Not only have they been using images as an accompaniment to movement for as long as the history of dance, but they also consider them to be effective as performance facilitators, whether in the creative, technical or educational setting (Loren 1978). Should we believe in the knowledge of the academic experts or in the experience and feelings of experienced dancers themselves? (Hanrahan and Salmela 1990: 18)

Sweigard herself was, I think, abundantly clear about this:

> 6. Imagined movement should be used both before and during voluntary movement. *There is always a tendency to start the actual movement too soon,* once it is understood, without taking time to concentrate first on the imagery being used. The sequence after the movement has been learned is (*a*) complete rest, (*b*) imagined action, (*c*) slow movement with continued *concentration on imagery,* (*d*) sudden release of effort, and (*e*) sensing the change, if any. (1974: 276—italics added)

Given that we accept Dr. Sweigard as an academic expert, the answer to Hanrahan and Salmela's question is this: The benefits of mental imagery on dance learning are maximized when time is taken to concentrate first on the imagery being used *without moving.* After the movement has been adequately visualized, the sequence outlined by Sweigard (Item 6 above) should be followed.

Sweigard had two additional messages about images and movement, which were: "[Item] 7. Voluntary movement must be *slow*" and "[Item] 8. Imagined action during voluntary movement of any part or parts of the body is invariably located in the trunk. Its direction is often in the opposite direction of the distortion the voluntary movement will produce in the trunk" (1974: 276).

I am somewhat perplexed by Hanrahan and Salmela's use of the words *mental imagery*—do they mean *body* imagery or *aesthetic* imagery? And what do we do with imagery that exists because it aids the dancer to construct a characterization, as in Lucas Hoving's case with the Christ figure in *Traitor*? Is there a difference in the images used by a professional dancer when he or she is working with the story line of a ballet and the characterization of a part, as against imagery typically used in a classroom, as Kirkland discovered? There are many questions regarding images, not all of which will be answered in this book, but further discussion in the future from scholars interested in the subject would certainly be desirable.

For now, suffice to say that Sweigardian ideokinesis aimed to change an individual's posture to an alignment that was more efficient. With reference to voluntary movement, students were meant to follow the steps outlined on page 67 (this volume). The procedures I used with professional dancers involved Sweigard's work on tension reduction in the CRP, use of the light beams, and the mental practices outlined at the end of Chapter 5.

In the meantime, we will get on with mirror images—a kind of image with which everyone is familiar but that presents serious problems in dance classes.

Appendix to Chapter 4
Better Dancing through Better Body Balance

LULU E. SWEIGARD

Among dancers, equally blessed with all other qualities which contribute to their success, and subjected to equally good teaching, the dancer with the best posture will progress faster, reach a higher degree of attainment, and last longer in his profession. In fact, there is probably no activity in which one's body alignment, or posture, is more important than in the dance.

The term *posture* is defined here as that body alignment which persists whenever an individual stands with his weight, as judged by him, distributed on both feet. There is only one posture pattern for each individual, and this must not be confused with body positions, of which there are many. One's body alignment is most stable and consistent in the central area of the body, including the pelvis, lumbar spine, and upper thighs. As the body sways in maintaining its balance in standing, the change in the pattern of alignment in relation to the line of gravity will be greater as distance of the center of gravity increases.

The posture pattern improves and becomes more efficient the more the skeleton itself conforms with principles of mechanical balance. Yet it must be recognized that the skeletal design, with its many relatively small parts which must support weight, is such that perfect mechanical balance in the upright position can never be attained. There will always be some muscle work. Ideally this muscle work should be confined mainly to a constantly changing reflex coordination of deep muscles, with large muscles free for work as needed in off-center positions and in movement. This coordination cannot be directed, though this is surely implied by those who believe posture can be improved by mere exercises and by holding parts of the body in supposedly good positions. Such "corrective" procedures may result in better looks, but both flexibility and efficiency have been sacrificed.

Three of the many reasons why the dancer's posture plays a significant part in his success are discussed here.

Posture and Flexibility. Posture and flexibility are interdependent. The clue to their interdependence lies in the varying degrees of muscle tone or tightness which ac-

---

Reprinted, by kind permission, from the *Journal of Health, Physical Education and Recreation* [JOHPER], May 1965: 22, 23, and 56.

companies one's body alignment. As everyone knows, muscle tone is increased by work. As the skeleton habitually deviates from good alignment, some muscles must proportionately and constantly work more than others to maintain equilibrium in the upright position. These muscles develop a greater degree of tone—more often than not a tightness which is maintained, even in rest. Thus a pattern of hypertonicity of muscles (a pattern) occurs in accordance with one's deviations from good alignment. Hypertonic muscles, when they act as antagonists in movement, do not lengthen as readily or as completely as muscles of normal tone, and they respond more quickly to the stretch reflex; the more they are pulled in movement, the tighter they become and the more resistant to movement. Hence they constantly restrict in one or more directions the range of movement which is allowed by the nature of joints and their surrounding ligaments. This is the reason why a poorly aligned body is limited in the flexibility which is so essential to the dancer. It follows also that more muscle power must be applied to overcome restriction in movement by tight muscles. This leads to muscular overdevelopment with distortion of body contour, especially in the thighs.

Body Mechanics in the Dance. Principles of mechanics of movement are not the major consideration in many of the techniques of movement of the dance. The dancer, always seeking new patterns of movement to express his ideas, cannot be limited to those which conform to good mechanics in the use of the body. This again emphasizes the need for good body alignment, with its innate values for freedom of movement.

A typical example of poor application of principles of mechanics is the turn-out of the feet, a technique more commonly used in ballet than in modern dance. Poor teaching of the turn-out of the feet, plus overemphasis on the degree of turn-out, can be the cause of many muscular problems in the dancer, especially should he start his training with a poor posture. Since the turn-out of the feet is such a fundamental technique, a brief analysis is in order.

The turn-out of the feet occurs at the thigh joints, though a slight contribution, under some circumstances, can be made by rotation at the knee. When the knee is straight, this rotation (in anatomical discussion referred to as the "screwing home" movement) has already occurred and the turn-out of the feet must be accomplished by outward rotation at the thigh joints. If not, disturbance of alignment at the knee and ankle joints and distortion in the feet will result [see the quotation of Kirkland in Chapter 4, pages 60–61].

Experience in dealing with problems of body alignment in the turn-out, both when the knee is straight and when the demi-plié is being performed, leads to two precautions for the dance teacher. It is wrong to tell the student to tighten in the buttocks to achieve a strong turn-out of the feet, even though it is true that outward rotators of the thighs are located here. Voluntary tightening of the buttocks always interferes with good alignment of the lumbar spine, pelvis, and thighs, with increased tightness of muscles of the low back, the extension of the thighs, and the hamstrings. Tightness in these muscles results in decreased mobility in the spine and restriction of inward rotation and flexion of the thigh joints. Accurate explanation and demonstration of the movement involved in the turn-out of the feet is important, but emphasis on muscle tightening in any part of the body is a highly questionable procedure—one that can interfere endlessly in efficient coordination of some 600 muscles throughout the body.

The other precaution relates to the manner of assuming fifth position of the feet. Sometimes the student bends the knees first, places the feet in fifth position, then straightens the knees. This is wrong. The fifth position should be assumed without bending the knees. Turning the feet out when the knees are bent means rotation occurs as usual and is not superimposed on the first rotation at the knees when they were bent. This combination of rotations places an undue amount of strain on the ligaments around the knees, especially those on the inside. Repetition of this method of assuming fifth position finally results in weakening of the knee ligaments. Since these ligaments are nonelastic, once they are lengthened by repeated stretching, they never return to their normal length and the knees lose one of their greatest protections against injury. The high incidence of knee injuries among dancers leads to the belief that perhaps skill and care in teaching the turn-out of the feet without distortion in body alignment is of fundamental importance to advancement in all techniques of the dance.

With prolonged poor training during early life, poor body alignment and muscular problems become so deeply ingrained that no amount of good teaching in later life can repair the damage done, especially at the knees, ankles, and feet.

Posture and Two-Joint Muscles in the Legs. Two-joint muscles of the lower limbs function either to extend the thigh, knee, and ankle joints, or to flex the thigh and knee but not the ankle (there is no two-joint flexor of the ankle joint). Extension or flexion in these joints seems to occur simultaneously, but it actually starts at the thigh joint, thus transferring the power of one-joint muscles at the pelvis to the most distant part of the limb.

The dancer is probably most aware of the two-joint hamstrings, because they restrict bending at the thigh joint when the knee is straight, as in the high kick, or in bending the trunk forward when sitting on the floor with straight knees. Restriction of this movement is to be expected. Even so, if body alignment is good, either one of these patterns of movement should be relatively easy. Only when movement is carried beyond normal range, as it often is, does the length of the hamstrings become a problem. For those with poor posture, the problem is much more difficult because the hamstrings are already tight in the standing position, before thigh flexion with straight knees is attempted. Under any circumstances, lengthening the hamstrings by forceful stretching results in lengthening of soft tissues, not, as it should, in elongation of the contractile muscle fibers. The latter is achieved mainly through attaining a better body alignment with its concomitant increase in degree of flexibility in joints.

Juilliard's Posture Laboratory

Our one aim in the posture laboratory at Juilliard is to help the dance student attain the best body balance possible for his work. Such help does not call for exercises as often advised by dance therapists. There must be recoordination of all muscles toward greater functional efficiency, especially those which maintain equilibrium of the trunk in the midst of the great variety of patterned movements used in the dance. Complete methods of teaching posture toward this end cannot be presented in this brief discussion, but two recommendations are given which may be easily followed by anyone seriously interested in improving his body for the dance.

The first deals with the use of the body off the dance floor. Good mechanics in everyday movement should become a habit. Above all, the dancer should walk with his toes straight ahead, not in the turned-out position used in dancing, and when standing, he should stand on both feet. The use of the body when not dancing makes its contribution toward building muscle habits—the discipline of the dancer really extends into all of life.

The second recommendation deals with the manner of resting. Every day the dancer should rest at least once (oftener if possible) on a level surface (not a bed) lying on his back with thigh and knee joints bent comfortably and arms lying across the chest. In view of the patterns of tight muscles occurring with poor body alignment, emphasized by dance techniques, there are precautions to be taken to make the rest position comfortable and free from the need of 'holding' any body parts in the prescribed position. The knees should be tied together just above the joints; the head should rest on a pillow of a height for comfort; a lift should be placed under the balls of the feet to keep them from sliding away from the pelvis; and when the arms and shoulders are uncomfortable, the elbows should be allowed to rest on the supporting surface with only the lower arms resting on the front of the body. If pain occurs in the low back or back of the pelvis, a small pillow should be placed under the lower part of the pelvis. In this position, the force of gravity does a great deal to ease and thwart increase in muscle tightness in various parts of the body.

The physical problems which tend to occur during the years of training of the dancer are many and varied, highly complicated, and sometimes extremely puzzling. Improvement in body balance does not make the dancer, but it does make the arduous task and long trek toward artistic perfection a bit safer and a little easier to take.

Chapter 5
Mirror, Mirror . . .

Mirrors made by polishing metal surfaces have been in use for centuries, beginning with ancient Egypt, where the aristocracy had gold, silver, and bronze mirrors as early as 2500 B.C.E. (Before the Common Era). However, mirrors as we now know them were first created in Venice, Italy, in the sixteenth century. They were made of glass with a backing of tin-and-mercury amalgam. Glass that is coated with a silver backing came into use in the nineteenth century, *circa* 1840.

Interesting as this is, it isn't mirrors themselves, their history, or their composition that are at issue in the following assessment of mirror images and their effects upon teaching dancing. It is the ways in which mirrors are used in dance classes that is the big problem. I also want to emphasize that it is the *mirror images students have of themselves* that generate challenging difficulties for teachers of dancing.

It may seem to some that *students* (not teachers) are at the greatest disadvantage with regard to using mirrors in dance classrooms, but 1. this is a book primarily about teaching dancing; therefore, I emphasize teaching problems, and 2. although teachers are assigned sole responsibility for having mirrors in their classrooms, I think mirrors are there on the whole because they are simply taken for granted: they are seen as standard equipment that few, if any, question. To my knowledge, not much thought has been given to their desirability or the consequences of using them. Readers may well ask why.

For a start, mirrors turn *four-dimensional* bodies (three dimensions of space and one of time) into *two-dimensional* images. The outstanding characteristic of any mirror image is that it is flat. Mirror images closely resemble paper dolls. Of the three spatial dimensions involved in space/time (front/back [F/B], right/left [R/L], and up/down [U/D]), the dimension that is commonly flattened in a mirror (and in students' minds) is that of F/B. The greatest distortion of dancers' bodily images occurs here. If the dancer has his or her body turned

Images 14 and 15. The greatest distortion of dancers' bodily images occurs here. If the dancer has his or her body turned with the right or left side to the mirror, then the flattened dimension is R/L (and the head must be turned in order to see the mirror image). © Human Kinetics with the right or left side to the mirror, then the flattened dimension is R/L (and the head must be turned in order to see the mirror image).

As everyone knows, it is nearly impossible to see the back surface of one's body in a mirror (which is why dressing rooms in department stores, for example, have at least three mirrors, permitting a back view of the body without requiring the twisting or turning the torso).

Two-dimensional Images versus Kinesthesia

In general, we can say that *any* mirror image of the body is an *a priori* reduction of the living human body into something that it is not. Unfortunately mirror images tend to become the dominant images of the body in nearly all dance students' minds. We will return to the subject of this kind of dominant image later on. For now, we will move to the mirror image and real movement distortions.

The chief technical problem that arises with actual movements in a dance-technique class is connected with the position of the head and its relationship to the rest of the body. In their efforts to watch their mirror-images, there is a tendency on the part of dance students to hold or "fix" the head in a relatively static position. To see the mirror image constantly throughout the class, the student holds his or her head in a stationary position, so that using a mirror has a direct, wholly deleterious effect upon the execution of dance moves themselves. Mabel Todd devoted several paragraphs to discussions of the head and its relationship to the rest of the body. It is important, I think, to reflect on head positioning at the outset, because it is in some sense an index to further problems.

1. This centering of the head is necessary, as the organs which report to us the position of the head in relation to the earth and keep us informed as to our place in space at any given time, are located immediately above the condyles. These sense organs are housed in the vestibule of the inner ear, and the ear is in line with the condyles (Todd 1937: 100).

2. *A head which is held habitually in an unbalanced position* [italics added], on one side or too far back, with the throat sagging, not only fatigues the muscles holding it, but must produce confusion in the proprioceptive mechanism.

   The head weighs from fifteen to twenty pounds. This weight must be balanced. If, while keeping the chest supported, you let the head hang forward of its own weight, you may realize how heavy it is (Todd 1937: 101).

3. The head is the top-load on the spine. When the main units below are in balance, it will rock easily on its support, cradled as it is upon the broad base of the atlas. It is further stabilized by the little pinion tooth of the axis, which holds and limits its motion. A little nodding motion like that of the mandarin doll, continued for a few moments, will give one a feeling of freedom in the neck muscles and make clearer to consciousness the fact of how deeply centered is the support of the head upon the spine. Gently up-reaching the head will increase the kinesthesia in this region, providing the forehead and chin hang in the same vertical plane. Getting a feeling for these pictures in the imagination will raise the head to a straighter position and bring it into line with the axis of the supporting cervical spine. (*Ibid*: 193)

From Todd's remarks, it is easy to see how a habitual imbalance of the head in relation to the rest of the body causes multiple problems throughout the rest of the body with regard to *everyday posture.* It needs to be said at this point that Todd's reference to everyday posture emphasizes the point of the head being in a habitually wrong position with reference to danced moves.

It reminds readers, yet again, of 1. the relationship of everyday posture to danced movements, and 2. the difference between imagery for the purpose of retraining everyday posture in contrast to imagery suitable for the teaching of dancing. Each student brings his or her everyday posture into the dance classroom. Working on everyday posture as Todd suggests will eventually improve conditions in dance classes.

Outside versus Inside Imagery

If, as is usually the case in a dance studio or rehearsal hall, mirrors cover an entire side of a rectangular room—a flaw that can only be changed by the teacher, who, knowingly or not, is teaching students that mirror images are more important than kinesthetic awareness simply by having the mirrors there—then 1. The students are thus moving from their habitual posture conformations, and 2. the body images they construct are visual in contrast to kinesthetic images. Visual images nearly always originate from a source *outside* the students' bodies, unlike kinesthetic imagery, which originates from *within* the students' bodies. Ultimately, of course, the visual images that any student has can only be changed by that individual student through becoming more aware of his or her own usage of imagery, making an effort to change what may be a misguided relationship to mirrors. The disadvantages of mirror use are not always obvious to students or teachers.

Keep in mind that the teacher is usually faced with a class full of students that he or she may not have trained from an early age, so that a predominance of *visual* mirror images in the students may be so ingrained that there is very little that can be done about it. Moreover, while it is true that teachers have a role in developing a mirror-image way of thinking about the body, we may well wonder what teachers who don't think that way are able to do in their classrooms when students can vote with their feet if they don't conform to the expected routines and environment of typical classrooms. It is well to remember, too, that many dance teachers consider their role *solely in terms of teaching dancing.* If the students' visual images are basically mirror images, if their postural patterns are poor, they do not feel responsible for these deficits. They will tell you that they are teachers of *dancing,* not dance therapists, or teachers of posture and movement, in any case.

Mirrors and Barre Work in a Ballet Class

Only the students facing the mirror will be able adequately to see themselves in the mirror from their positions at the *barre.* Students who have their right or left sides to the mirror can only see their mirror image by turning their heads to the right or left, which is *not* the correct position for the head during most *barre* exercises. Moreover, an interesting process of internal talking by the student to him- or herself often takes place while the student looks at the mirror image, so that concentration is directed *away* from the movement to the (unvoiced) criticisms of his or her appearance and related matters.

The student becomes a fascinated spectator of the movements made by his or her mirror image, *not* a monitor of his or her own movements from a kinesthetic point of view. There is a sense in which students become spectators of (not creators and controllers of) the moves they make throughout the entire process.

There is nothing abstruse or hard to understand about this. In common with everyone else, dance students have mental pictures, both of themselves and (perhaps more important) what they *wish* themselves to look like. Generally, their ideal images strongly contrast with what they see (or what they think they see) in a mirror.

Internal Talking

In addition, after teaching for years, I became convinced that the constant stream of internal talking that takes place on students' part while looking at their images in a mirror distracts them from full class participation, because they tend to hear very little of what the teacher says. Their interaction with other students is inhibited, and their private worlds become more real than immediate classroom events.

Many, if not all, dance students engage in an essentially private process of checking, revising, wishing, correcting, and comparing their mirror image to partake of some "ideal" they have formed. Because of this process, corrections or observations made by the teacher are often (although not always) unheard. Students' attention is so focused on the stream of internal talking that sometimes, they literally do not hear anything—even the music.

Teaching classes or directing rehearsals in a mirrored room thus means that teachers are faced, first, with a battle for students' attention. Then, too, the students' internal assessments of their mirror images may be at odds with reality. It is well known, for example, that anorexic female students see themselves as fat and that slender male students think they are not masculine enough, and so on. There are uncountable numbers of possible negative assessments possible in the relationship between individual students and their mirror image (see Appendix 5 for another aspect of negativity that often characterizes the dance classroom).

The situation is worsened when center floor work is undertaken. Most of the time throughout a ballet class, for example, the head should be in such a position that the dancer's gaze follows the direction of the movement, either of a bodily part (often, the arms), or in a designated direction through space.

For instance, an arabesque, performed side-on to the mirrored wall, usually requires the dancer's gaze (eye-line) to follow the line of the forward arm, which means that the dancer *cannot see* the mirror but must look at one of the blank walls of the room (see Images 14 and 15).

The temptation to peek—to look at the line of the arabesque in the mirror—is well nigh irresistible. To see a mirror image, the student's head must be turned *toward* the mirror, with the result that the line of the arabesque as a whole is destroyed. Entering the realm of characterization, mirrors are a disaster, first, because there are no mirrors onstage. There, dancers must rely on their kinesthetic awareness of their own bodies and the bodies of others with whom they are dancing. Eye-lines and such must follow *characterizations,* which follow the story line (or lack of one) in a specific dance. Thus mirror images become liabilities in more ways than one.

Similarly, when a series of turns or jumps is undertaken along a diagonal line through space, the eyes should be directed toward the line of movement, not at the mirror image of the starting position. It is impossible, in any case, for a dancer to see him- or herself while performing a series of fast turns across the classroom.

I won't provide further examples, as I assume that teachers (and students) can supply many examples of their own; however, I want to emphasize that these aspects of teaching dancing are both inaudible and invisible. The teacher cannot hear or see the negative evaluations of themselves made by the students that arise from the images they form of themselves through taking technique classes.

Gelsey Kirkland says it very well: "As in the myth of Narcissus, the beautiful youth who falls in love with his own reflection, the relationship between the dancer and her [or his] mirror image is an intimacy of extraordinary power and potentially perilous consequence. Most dancers ultimately seem to drown themselves in their own images, pushed by forces unseen. The dimensions of tragedy are revealed only when lives and personalities are destroyed. Until then the damage remains invisible" (Kirkland 1986: 79).

The internal talking that takes place occurs within the silences of each student's own self. Such internal conversations as students have are rarely shared, but they form the basis for the students' assessments of what they are doing, of the success of the classes they take, and of their overall achievement. Needless to say, perhaps, the preoccupation with their mirror image doesn't assist them in any real way when they dance onstage, where there is no mirror—only the well-known "black hole."

At this point, it seems vital to say that there are many issues regarding this subject that merit further examination, but I must be guided by the aims and objectives of this book—to clarify the most important usages of "imagery" with reference to dance *teaching*. What I hope to achieve is a basis for thoughtful discussion in written publications and departmental seminars about dancing in the future, such that the unexamined use of mirrors in classrooms might change.

Unpopular Comments

When teaching master classes outside my own studio in mirrored classrooms, I gained a reputation for making acerbic comments that were not often appreciated: "Stop playing Alice Through the Looking-Glass," I would say, or, "If you are really as ugly as you think you are, then why are you trying to dance at all?" These statements were disconcerting to students, gaining for me a reputation as mind reader or, less charitably, a crank. I often wondered how students could fail to recognize the transparency of their facial expressions when they were engaged in their routines of internal talking and negativity. Again, Kirkland says it very well:

> Throughout the early phases of my career, the mirror was my nemesis, seductive to the point of addiction. Stepping through the looking glass meant confronting a double who exposed all of my flaws and pointed out all my physical imperfections. Over a period of time, the image in my mind meshed with the image in the glass. Until the opposition between the images was resolved, I saw myself as a walking apology, unable to attain or maintain my constantly refined ideal of physical beauty.
>
> With all of my insecurities intensified, I became my own worst critic, embarking on an aesthetic quest for perfection that in the end would heal the wounds I had inflicted upon myself. Trying to perfect both my appearance and the quality of my movement, I was unaware of a contradiction. As I continued to educate myself, my love and anger wed within my personality. I remained a child. I worked and lived in isolation, an almost absolute solitude that I now see was unnecessary. I was misled, sometimes deliberately. And I was not alone.
>
> The endless repetition of *barre* exercises in front of a mirror reflects a distorted image many people have of ballet, an image shared by many dancers. (Kirkland 1986: 79)

Kirkland is aware that "few dancers have taken the relationship (with the mirror image) to the extremes of [her] early career," but "fewer still have managed to reverse the hold of the mirror" (1986: 73).

It has been drawn to my attention that some postmodern dance practices (e.g., contact improvisation and release work) are never taught in mirrored environments and that this has been happening since the early seventies—a commendable practice. I've also been told that in some universities (Arizona State among them), mirrors are now draped, so that they can be used more consciously. I would hope that this becomes standard practice. It would certainly improve the teaching and learning of any form of dancing.

Should my discussion so far of teaching with mirrors be distasteful to some readers, then I am sorry, *not* for what I have said, but for those who resist the profound implications that the injudicious usage of mirrors has on the teaching and learning of dancing—*any* form of dancing. It is gratifying to know that recent academic research agrees with my point of view (see Radell, Adame, and Cole 2004). Unfortunately, I haven't the time or space to review this paper here, but I hope readers will seek it out, for it is an important milestone in pedagogy.

To summarize: If readers can agree that dance students need to learn to protect themselves from poor *movement* imagery in a dance class, and that they must also learn to protect themselves from the powerful seductions of mirror images and the misguided self-criticism that these nearly always imply, then they will have helped me to achieve the goals of this writing. Like

Kirkland, "I suspect that every dancer experiences the mirror in a uniquely personal way, although many are perhaps oblivious to the power it exerts over their lives" (1986: 79). She attaches the power of the mirror image to a "state of artistic servitude" and how the state of servitude comes into being:

> As a primary teaching tool for dance, the mirror fosters the delusion that beauty is only skin-deep, that truth is found only in the plasticity of movement. *It seems preferable to imitate rather than create.* Imitation can be varied to create the impression of originality. There are endless possibilities for breaking the human mold into novel patterns. To be daring in dance no longer involves risk, virtuosity and strength of conviction. The dancer can win approval for steps that require no real decision in creative or compositional terms.
>
> The dancer is trained to *watch,* to enter the world of the mirror until it is no longer necessary [even to] look. To the extent that a dancer becomes a complacent reflection, he or she does not learn how to test beauty, how to discover its inner life. In this way, the mirror can trap a dancer's soul, ultimately breaking creative spirit. Such a dancer is created, but does not know how to create. With success and popularity, the situation becomes more precarious. At any moment, with the capricious changes of fashion, a glance to check the mirror may reveal tragedy—that he or she has been created for nothing. (Kirkland 1986: 80)

All of the above furnishes the reasons why I did not install mirrors in my studio in New York, nor use mirrors in my studio in Portland, Oregon, where I began teaching in the late forties. I always had one or two ordinary full-length mirrors, but they were draped. They were only used occasionally for specific purposes pertaining to skeletal alignment.

Not using mirrors has never been a popular stance to maintain in the dance world, and I suspect that it is no more popular now than it was fifty-odd years ago; however, I hope I have provided ample reasons why the practice should be abolished—at least to the extent of draping the mirrors so they can't be used all the time. Another form of protection for students is occasionally to have short discussions of mirror images and what they can (or might) mean to them.

Kinesthesia and the "Five"—(Really Six)—Senses

To recapitulate: the use of mirrors inhibits and, in some cases, destroys the development of kinesthetic imagery with reference to any movement form. In an attempt to understand the difference between kinesthetic imagery and mirror imagery, we will start with a particularly useful form of it called "mental practice"; however, before we begin the discussion of kinesthetic imagery, some general points seem necessary.

With our discussion of mirror images complete, we've reached the end of an easy definition of the word *image.* Everyone knows what a mirror image is, but not everyone knows that images with regard to movement differ markedly in kind. Here, we enter the world of different sensory modalities, usually defined by what we are all taught as the five senses, that is, touch (skin), taste (tongue), smell (nose), hearing (ears), and vision (eyes). *The most important image for the teaching of dancing (sports, etc.) is left out:* the kinesthetic image. It is again important *precisely* to define what is meant by *kinesthesia,* thus we will look again at a quotation found in Chapter 3, page 41:

> Notably absent from this conventional taxonomy, however, is kinesthesia, our sensory awareness of the position and movement of the body. I ask the reader to please close your eyes and lift your arm—move it around and ask yourself how you know where your arm is located? This is kinesthesia; literally 'movement' (kinetic) + 'sensitivity' (aesthesia). It is this kinesthetic sense that provides information on the whole repertory of our motor actions, from the raising of an arm, to walking, even to the turn of the eyeballs and swallowing. Physiologically speaking (that is, in the discourse of Western natural sciences) kinesthetic sensations are registered by receptors in the muscles, tendons and joints of the body. As the muscles function when we move bodily parts, various patterns of pressures on these receptors provide essential information for the guiding of motor action.
>
> The perception of spatial movement and orientation of the body as a whole also involves a fluid-filled receptor system located in the vestibules of the inner ear. More than balance, this structure provides the means by which we are aware of being tilted, shaken, or whirled about, and how, most of the time, we know "which way is up"!
>
> The exclusion of kinesthesia from the Western taxonomy of the senses—this (ab)sense, as it were—is particularly interesting because scholars of perception as diverse as Descartes, Dewey, Gibson, and Merleau-Ponty all acknowledge body movement as the unexamined ground of all sensory perception. One is led to ask why, then, has kinesthesia been excluded from consideration? (Farnell 2003: 133–34)

This sixth sense of kinesthesia provides us with experiential images of what it is to walk, run, fall, balance—all of the actions we perform every day—and it provides us with our fundamental sense of where we are located in physical and geographical space.

When I said earlier that "the use of mirrors inhibits and, in some cases, destroys the development of kinesthetic imagery with reference to any movement form," I was not being flippant, nor do I exaggerate. Focusing on mirror images (the sensory modality of *vision*) in a dance classroom *does* inhibit the development of kinesthetic imagery (the sensory modality of sensing through the joint ends and inner ear of *kinesthesia*) by focusing the student's attention on what can be seen *outside* of the body rather than what is experienced *internal* to his or her body.

The Complexity of the Concept of Image

Students should also be aware that there are eight primary usages of the word *image* in most English dictionaries (see, for example, *Webster's Third International Dictionary,* 1961: 1128—Vol. 2). Of these, only three definitions are relevant to the use made of the word *image* in this book. We have already explored definition 2 given in *Webster's International* in great detail:

> A thing actually or seemingly reproducing another as an optical counterpart, i.e. "mirror image," or the image produced by an optical instrument such as a lens. This usage can include such phenomena as photographs [films, video recordings] and "acoustic images," as in the reproduction, by mechanical means (tapes, CDs), of the "image" of a person's voice.

The third set of meanings under the fifth definition of *image* in *Webster's International* leads to "idea" or "concept"—to the notion of "a mental picture." This is a good definition of *image* for dancers and bears the meanings typically found in Sweigardian ideokinesis.*

Mental pictures of this kind are incorporated in the discipline of "mental practice," best understood, I think, by examining an early psychologist's work in this area (see Richardson 1967a, 1967b, and 1969).

Mental Practice

Richardson's definition (and use) of imagery incorporates traditional criteria used in academic psychology by which *percepts* (reality) and *images* (fantasy) have been distinguished, pointing out that the differences are not now (and have never been) absolute. He introduces the notions of perception, memory, and thought, observing that perception and remembrance can be carried out without any self-conscious awareness of the objects perceived or remembered, using the example of driving a motorcar while recalling the events of a football match (Richardson 1969: 11). It is when he talks about "mental

*Remembering, too, that Sweigard's practice of ideokinesis included *tactile sensations* such as those described in Chapter 1 with regard to light-beam images.

practice," however, that he makes a real contribution to the consideration of imagery and dancing.

He defines *mental practice* as "the symbolic rehearsal of a physical activity in the absence of any gross muscular movements," giving as an example a high jumper who *sees and feels* himself going through the run-up, takeoff, roll-over, and landing required for that type of jump. A high jumper who does this sort of thing *with the conscious intention of improving his performance* is engaged in the discipline of mental practice. Richardson further declares that "improved performance can result from this form of practice and . . . visual and kinesthetic imaging abilities may well play a significant role in the amount of gain obtained by any particular individual" (1969: 56).

In mental practice, the dancer or athlete *is not* visualizing a mirror image of him- or herself performing imagined actions—as if watching a film clip. The person sees him or herself *behind* the image of the body, or actually "in" the image. The clearest picture is that of the movement-writer (or Labanotator): when writing movement, the writer must see the movement from behind, as it were—-walking, dancing, or whatever from the performer's point of view, not that of the audience.

The high jumper senses and feels him- or herself poised at the beginning of the run-up to the high jump, then the run, the takeoff, the roll-over, and the landing. The whole sequence is therefore *sensed kinesthetically* by the individual in terms of his or her own body actually performing the sequence of moves, but none of the moves are actually performed. In other words, the image is a *kinesthetic image* that arises directly from the individual's *kinesthetic memory* of the act. Richardson says,

> The significance of imagery can be only in our awareness of its presence and of what we are able to 'read' from it. It is a basic assumption that imagery is not merely an epiphenomenon but that one's conscious awareness of its presence may make a difference in our behaviour. That being consciously aware of something has implications for behaviour, is an assertion that should be obvious but, like the study of imagery itself, it is only in the last decade [1959–69] that concern with the general issue of consciousness has been revived. (Richardson 1969: 143–44)

Behind an Historical Eight Ball

It is important now, I think, for readers to realize how early in the twentieth century Sweigard's work was carried out. Why? Because there was a gap in academic psychology, as Richardson points out, created by the growth of behaviorism in the 1920s, where "mental imagery began to fade as a serious subject for investigation" and where it "completely disappeared as a theoretical construct," although "From the 1950s academic psychology has expanded the range of research into areas concerning cognitive processes," of which the activity of imaging is one" (1969: ix).

Few people seem to realize that Sweigard's original work was done in an extremely inhospitable academic atmosphere. There is no doubt that it was scientific, because of the theory, language, and methods she used, but her work involved the use of imagery and consciousness, both of which were seen as liabilities. Because of this, she resisted any incursion into her work of anything she believed to be unscientific. She was truly afraid that her work would be misused, "and the result of her refusal to codify her work has been that much of what she taught single students that was meant to be applied only to specific problems is now being taught as the whole of ideokinesis" (Williams 1982: 9).

Sweigard taught in an historical period characterized by strong divisions between that which was mental and that which was real. Richardson remarks: "Memory imagery is the common and relatively familiar imagery of everyday life," quoting William James to clarify:

> Mental fire is what won't burn real sticks; mental water is what won't necessarily (though of course it may) put out even a mental fire . . . . With 'real' objects, on the contrary, consequences always accrue; and thus the real experiences get sifted from the mental ones, the things from our thoughts of them, fanciful or true, and precipitated together as a stable part of the whole experience-chaos, under the name of the physical world. (James 1912: no page number given, cited in Richardson 1969: 43)

With regard to mental practice, Richardson declares that mental imagery consists of "all those quasi-perceptual experiences of which we are self-consciously aware, which exist for us in the absence of those stimulus conditions that are known to produce their genuine sensory or perceptual counterparts" (1969: 2–3).

Moreover, having, I trust, established a crucial distinction between visual and kinesthetic imagery in this chapter, we will move on in the next chapter to consider the characteristics of aesthetic imagery, kinesthetic memory, and other matters—all of which are connected with our habitual movement practices, whether in everyday life or in dancing.

Appendix to Chapter 5

Accentuate the Positive . . .

Perhaps few people today will remember an old popular song that began: "AC-CEN-tu-ate the positive, E-LIM-in-ate the NEG-a-tive. . . DON'T MESS WITH MR. IN-BETWEEN," but I recall these words well—though nothing beyond them has remained.

Not only was the song in vogue when I began studying ballet in the early nineteen-forties, it provided an idea about teaching that I never forgot. Although the song-source for the idea may seem trivial, the message it contains for the use of imagery by teachers of dancing is profound.

Many teachers of dancing are guilty of a common error in using (more accurately, misusing) imagery in classrooms by constantly telling students what *not* to do instead of telling them what *to* do. My first ballet teacher was a shining example: looking back, I doubt that he ever gave a positive instruction when he taught dancing.

To illustrate first with an example outside the dance classroom: suppose someone says to you, "Don't think of the Empire State Building."

What are you thinking of—what is the *mental picture* created in your mind?

A mental picture of the Empire State Building, of course!

If (for whatever reason) that person doesn't want you to think of the Empire State Building, the thing for him (or her) to do is to tell you to think of the Chrysler Building, the Rocky Mountains (or whatever) instead.

This simple rule applies to any image pertaining to dance classes or dance performance whatsoever. Even at age thirteen, in beginning ballet classes, I found myself continually having to figure out what I was *meant to do* from that which I was told *not* to do!

In dance classrooms, teachers often say, "Don't bend your knees." The image they instantaneously create in students' minds is that of *bent knees;* thus the teacher's use of imagery creates an image of the mistake. The image serves to deepen (not lessen) the student's awareness of a specific blunder connected with a specified move, rather than heighten the student's awareness of what it is that he or she is meant to do.

If a dance teacher doesn't want bent knees in a specific movement, then why doesn't he or she say, "Straighten your knees"?

Granting the fact that focus on the desired movement outcome (rather than the mistake) is difficult for the teacher to accomplish, especially if he or she has taught by means of negatives for years, it is nevertheless achievable. The effects of applying this simple rule throughout an entire class are often astonishing.

The lesson here is that the use of imagery in dance classes has unambiguous negative (and possible positive) consequences. Thus we may say that images that are body-directed—that is, *any image that names a bodily part,* combined with a direction (as in the example above, knees and bend), should be stated in positive terms. That is,

"*Straighten your knees*" instead of "Don't bend your knees"
"*Eyes straight ahead*" instead of "Don't look at the mirror"
"*Push the floor down*" instead of "Don't lift yourself up"
"*Raise your arm shoulder height*" instead of "Don't raise your arm so high"
"*Bend forward from the thigh joints*" instead of "Don't bend from the waist"
"*Rotate thigh joints outward*" instead of "Turn out your feet"
"*Extend your ankle*" instead of "Point your toes"
"*Stretch your toes*" instead of "Point your toes"

This list could be extended *ad infinitum.* But someone says, "What about giving *wrong* directions positively?"

Okay—to give a wrong direction positively is not good; that is, "Tuck your pelvis under" is a positive verbal language statement that creates an image of something teachers emphatically should *not* want their students to act upon or to perform. Here, the corrective isn't quite so simple, but images that serve to lengthen the spine downward, or to direct the student's attention toward widening the pelvis horizontally across the back, can serve to eliminate the image of tightened buttocks.

The same can be said of "Turn out your *feet*" and "Turn out your *knees*": here, the students' attention should be focused on the thigh joints, not the feet or knees. Similarly, "Point your *toes*" needs positive focus on extending the ankle and/or stretching the toes. It is difficult to make general statements about such images as "Keep your *shoulders* back" and "Lift your *chest,*" because a caved-in chest may be the result of what I call the "Vogue model syndrome" and may be a postural habit that requires more than single (or multiple) images in class to correct, or it may be a misunderstanding on the student's part of where his or her arms are meant to be—or any one of a number of other things too numerous to mention here.

Suffice to say that the creative use of imagery in a dance class should be based on a solid background of functional anatomical knowledge, thoughtfully applied to specific circumstances and to the specific bodies with whom one is working. In other words, teachers of dancing should carefully examine his or her classroom imagery that names specific bodily parts (as in the above cases), taking appropriate steps toward correction if necessary.

Do Images Always Communicate?

There are other problems with imagery in dancing classes, clearly illustrated by an anecdote supplied by Carol Knopf (New York University) about imagery in a ballet class. She says that her first consciously received exposure to imagery came from her ballet teacher. She was twelve. The image was: "Arabesque as if you have a lemon in your stomach."

> I thought it sounded a bit odd, but I was too intimidated by my elegant instructor to ask her what it meant. I went home and thought about the image, deciding that it must mean that when you assume an arabesque position, you imagine that your body, from the top of your head to the toes of your raised back leg, must be like the gentle curve of the side of a lemon. To arch into an arabesque like the side [outline] of a lemon, eliminates the error of jamming your lower back by trying to keep it completely straight. It also makes a beautiful long line. Almost every day my instructor said, "Remember the lemon!" For the rest of that year, I always used the lemon-in-the-stomach image. My arabesque continuously improved. It felt better. "It's longer and stronger," the teacher declared.
>
> After many months of success in her class, I felt comfortable enough in the teacher's presence to thank her for the lemon image, which had helped me so much. "So!" she said in a shocked voice, "All this time you have not understood me! When you eat a lemon, the tartness makes your lips contract. This is the action I want from your stomach muscles when you arabesque!"
>
> Feeling as if I had suddenly been stranded in the middle of the Pacific Ocean, I spoke faster than I thought—as if dog-paddling toward a mirage of land—revealing my misinterpretations of several other images she had given, but, in class the next day, even though my imaged understanding was not hers, my physical responses evoked her approval.

This anecdote reveals several problems regarding the use of imagery in the dance classroom.

1. *Interpretation of the image:* Dance teachers can never be sure that the import of an image they give is going to be understood in the manner intended. The lemon image *to the teacher* aimed to accomplish contracted stomach muscles. The lemon image to the student evoked the image of a long line.

2. *Properties of the image:* The main feature of a lemon to the teacher was its tart taste and contraction of the lip muscles that she wanted transferred to the stomach muscles. The main feature of a (two-dimensional) image of a lemon to the student was a picture of the gentle curve of a lemon's outline, as in a line drawing.

3. *Assessment:* The teacher's image was absurd from an anatomical point of view. To contract the stomach muscles (located where—toward the upper left side of the body under the rib cage, or lower down, at the top of the pelvis?) would be contrary to the fact that the muscles of the front of the body must necessarily stretch (not contract) to allow for the raised back leg.

> The student's image, two-dimensional though it was, at least related in some sense to the *line* of an arabesque position, thus it was a more effective image than that of the teacher—but the use of light-beam images would have been much more effective.
>
> Both Carol's and her teacher's images of the arabesque pertain to *outer* surface(s) of the body—the teacher's directed toward muscles, which is wrong in any case. Carol's image was much better than her teacher's but could have been improved—that is, she was still thinking of *outlines* (not *inner* lines) of her body.

The difficulty in communicating and the use of imagery is that we tend to think of dancing as "nonverbal." Sufficient attention has not been paid to the complexities of the kinds and character of imagery or to the role images play in learning to dance, to excel in a sport, or, for that matter, in our everyday lives.

Chapter 6
Imagery and Habit

In the last chapter, I argued that there are different *kinds* of imagery important to the teaching of dancing. I focused on an important distinction between mirror imagery (outside the body) and kinesthetic imagery (internalized imagery) inside the body. Of the three definitions of image we began with, we have dealt with two: 1. mirror images and 2. the notion of idea or mental picture. The third definition of the word image in *Webster's International* is the following:

> 7. Something concrete or abstract introduced (as in a poem or speech) to represent something else which it strikingly resembles or suggests (as in the use of 'sleep' for 'death'). Compare EMBLEM, SYMBOL. Or a figure of speech (as a metaphor or simile—a 'trope').

Here is a definition of *image as metaphor* frequently used in dance classes. This is the most common use of images with reference to the aesthetics of dancing, where students are encouraged to move "like" something else or "as if" something else.

Keen students will notice that Dr. Sweigard also uses metaphors with reference to anatomical imagery, but we will talk about that later. I arbitrarily use *metaphor* to distinguish aesthetic from bodily imagery, in the following table. It should be clear by now that my usage of anatomical imagery means *using imagery while the physical body is not moving.* It should also be clear from the 'thought exercise' to follow that Sweigard's use of metaphors for anatomical parts arose because many of her students were not able to visualize the actual bodily parts to which the image was attached—for example, a turtle in place of the distal end of the first metatarsal bone of the foot.

For the sake of clarity, I have devised a brief exercise in images for readers that is meant to elucidate the differences between anatomical and aesthetic imagery. Sweigard's use of anatomical imagery in the CRP forms the basis of

Anatomical	"Mental Practice"	"Aesthetic"	Mirror
1. Used in Ideokinesis sessions, *with tactile sensations supplied by a teacher,* utilizing *no* outward movement, meant to change the subcortical pattern of response with reference to posture.	1. Images of *whole sequences* of moves (i.e., high jump, a passage in a dance, a vault in gymnastics, a particular tennis swing, etc.) also visualized internally *without moving* the body.	1. Used in dance classes to aid performers in *producing* movement. Generally metaphorical, meant to improve the appearance of the performer's moves with reference to desired aesthetic qualities of specific moves.	1. A *reversed* image of a person, seen in a mirror, that is apart from the individual's moving body. This image is *not* connected with a kinesthetic, internalized, image.
2. Used in dance classes connected with standard verbal labels, i.e., 'thigh joint,' 'rib cage,' and such. Meant to locate specific points or areas within the body connected with specific moves.	2. These are internalized (kinesthetic) images of the whole body, meant to increase precision and accuracy of a sequence in performance. Similar to images necessary to movement-writing.	2. Self-made images of dancers constructed to help them perform, such as "Push the floor down" to achieve lightness and a "floating" quality. These are also internalized or kinesthetic images.	2. These images are often derived from a person's self- assessment that tend to be one-dimensional, masking much more than they show. In some sense, they are the kinds of images an audience has.

the exercise: she used the 'turtle,' meant to shorten the distance between the big toe and the heel (see Sweigard 1974: 243–44). How different is this from many aesthetic images of the foot (or feet) that may occur in a dance class, such as, say, "a bird flying through the air" or the foot "fluttering against the supporting ankle like the rapid beating of a heart"?

The first time (and every time) I encountered Sweigard's image of the turtle,* it was accompanied by the pressure of her right thumb and index finger on the top and underneath the distal end of the first metatarsal and the proximal end of the phalange of my big toe on my right (or left) foot. The turtle crawled from the toe to the heel (never the other direction), allowing the subcortical processes in my body to make suitable neuromuscular adjustments between the big toe and heel of the right (or left) foot.

While I visualized the turtle, my entire body was in a state of reduced tension. The pressure of her fingers *located* the image precisely where it was meant to be. There was *no movement involved* except on the part of the imagined turtle. Here, I was reeducating a part of my body—going through the same process as I had with all the images of tension reduction (the suit of clothes, etc.) that she chose to give me during that particular session.

In general, anatomical imagery is meant to have some effect on 1. the habitual posture and habits of an individual; 2. more effective, efficient ways of performing a danced move—for example, the use of light beams. By way of contrast, aesthetic imagery is meant to have some effect on how a whole danced move is performed. For me, the use of anatomical imagery always worked to improve the postural habits of an individual so that his or her dance technique (whatever it was) seemed automatically to improve. Aesthetic imagery was meant to assist the dancer better to convey the emotional and psychological characteristics of a specified role or passage of movement in a dance which was (in turn) meant to increase an audience's appreciation of the whole dance.

Mental practice, on the other hand, is always about a *sequence* of moves—from a dance or from a technique class—that is visualized by an individual in order to improve his or her performance of the sequence. Richardson's example was that of a high jumper's progression from the start to the end of a whole jump. The important point here is that the sequence is *not* visualized

---

*I use the example of the turtle because Dr. Sweigard told me that (having studied bone anatomy) I could imagine the end of the bone moving toward the heel if I felt comfortable with that—use of the turtle wasn't necessary except for those who didn't know the shapes of bones. For many students, the turtle is not only easier to imagine than bone ends, it is the only option, if they haven't studied bone anatomy in depth.

as if the individual were separate from him- or herself, as in a mirror image. The imagery is, as it were, internalized with regard to his or her own body. It is kinesthetically *felt* (that is, sensed).

Unlike a mirror image, an *internalized* image—a kinesthetic image—is really the same, or very similar to, that used by a movement-writer (a Labanotator) in writing a dance or some other movement system. That is, the writer must see him- or herself *behind* the performer *as if* someone else's performance is his or her own—or internalized, if it is personal performance being written. Keep in mind that often, movement-writers deliberately use visual imagery in movement-writing for studying the choreographic design of the dance or the structure of a sign language. With regard to a whole choreography, the writer will often depend upon visual imagery to determine the distances and spatial patterns between (or among) performers and such. The important point is the shifts between visual and kinesthetic imagery. The writer closely identifies him- or herself with the moving body of a performer when writing his or her *movements.* The image in some sense *becomes* the writer if it is mental performance that is being done.

Dancing (the Act) and Imagery

In my classes, I used aesthetic imagery when I worked with students on performance or during center floor work in a dance technique class. I almost never used mirror imagery, and hoped that my students didn't. Undoubtedly, the subject of imagery and dancing is more complex than the majority of people imagine, but regarding the subject, we can now ask, "What is the relationship between imagery and dancing?" The question is significant because it raises a hotly debated issue regarding the use of imagery while the body is *passive,* in contrast to using imagery while the body is *moving* (see Hanrahan and Salmela 1990).

> According to specialists in the area of skill acquisition (Suinn 1976), the benefits of mental imagery on dance learning or performance are maximized when the dancer remains perfectly still. Dance experts, however, beg to differ (Dowd 1981). Not only have they been using images as an accompaniment to movement for as long as the history of dance, but they also consider them to be effective as performance facilitators, whether in the creative, technical or educational setting (Loren 1978). Should we believe in the knowledge of the academic experts or in the experience and feelings of experienced dancers themselves? (Hanrahan and Salmela 1990: 18)

In answer to the Hanrahan and Salmela quote, I would want to say that the benefits of anatomical imagery in ideokinesis are maximized when dancers remain perfectly still. The same is true for mental practice, but it doesn't hold for the use of aesthetic imagery. I would agree with Dowd that aesthetic images are "effective as performance facilitators" but would hesitate to commit myself further than that, for the reasons I've explained in previous chapters of this book. Imagery with reference to dancing is not a simple, unproblematic concept, as I've tried to show.

In my experience, out-of-context generalizations are nearly always wrong. By this, I mean to say that imagery is *not* a heterogeneous "thing" that can be treated always and everywhere in the same ways. In answer to Hanrahan and Salmela's last question ("Should we believe in the knowledge of the academic experts or in the experience and feelings of experienced dancers themselves?"), I would first ask, why place knowledge and feelings in an adversarial relationship? Why polarize the issue? I would then ask, *which* "academic experts" and *which* dancers—for this creates problems where none need exist.

In this book, I have depended upon Sweigard and Robinson for knowledge regarding anatomical imagery and mental practice. I have depended upon Gelsey Kirkland for her knowledge (based upon her feelings and experience) regarding aesthetic imagery. I have also depended upon my own knowledge and experience with performing and teaching dancing.

In my opinion, Hanrahan and Salmela are correct in pointing out that "While physical passivity may be a condition for ideal effectiveness of imagery, this does not preclude the use of imagery while moving" (1990: 18). Their work is useful because it points to the many different definitions (and usages) of mental imagery (cf. Richardson 1969). However, not all definitions and usages of imagery are compatible with regard to movement study, as we shall see.

"Creative Visualization," "Relaxation Meditation," and "Visioning"

The usages of ideokinesis, mental practice, and aesthetic imagery require rigorous self-discipline and concentration. Like ideokinetic imagery, the uses of mental practice and aesthetic imagery so far discussed are different because they are *specific to* posture, movement, and dancing, in contrast to different, popularized definitions and usages of visual imagery that can be found in magazines or on the Internet. For example, 1. <http://www.creativelivinginc.com>, headed by Karim Hajee, and 2. <http://relaxationmeditation.com/imagery.html>, headed by Michael Patrick Bovene.* The latter website's au-

*These websites may not exist anymore, but I'm sure that others can be found. Put the word *visualization* into Google and any number of things may crop up! All such websites have common characteristics.

thor (item 2) has written many articles on meditation published by Circle of Light Publishing from 1997 to 2006. The former (item 1) offers several courses designed variously to create wealth, set goals, encourage self-confidence, and much more, based upon "unleashing an individual's subconscious mind power" through visualization. The emphasis in both is on imagination in its broadest definition, and visualization with no restriction whatsoever. Critics of articles and courses like these have (rightly) spoken in terms of "pie-in-the-sky" goals and ambitions.

On the other hand, Courtney Helgoe (2006: 88–90) uses the word *visioning* instead of *visualization,* saying that "Essentially, visioning is a little like yoga for your brain cells—a chance for your mind to stretch and re-form itself around the experiences you'd like to have, thus creating the mental space and emotional energy for them to happen in actuality" (2006: 88). She goes on to say that, at a practical level, visioning is "a lot more than just vaguely and passively fantasizing about what you want. . . . Brain researchers now know, from observing positron emission tomography (PET) scans of people imagining or watching various scenes, that the brain reacts very similarly to real, watched and imagined experiences. At a neurological level, it [the brain] doesn't really know the difference. The brain also develops neural pathways that reflect frequent-use patterns, and builds synaptic connections that support habitual trains of thought. So the idea is that, by visualizing a positive experience intensely and regularly enough, you may be able to develop a mental and emotional infrastructure that can support and process it in reality" (*Ibid.* p. 88). Like popularizers everywhere, Helgoe uses authorities and presumed scientific facts to support her notion of visioning:

> Steven Ungerleider, PhD, leading sports psychologist and author of *Mental Training for Peak Performance: Top Athletes Reveal the Mind Exercises They Use to Excel* (Rodale Press, 2005), calls it "mental blueprinting." It works well for many athletes, he notes, because it helps develop the appropriate motor programs in our nervous systems: "Rehearsal of the sequence of movements involved in a task allows us to learn them symbolically," he writes. "We can then apply them when we go out on the field of competition."
>
> Ungerleider also points to a possible psychoneuromuscular reaction: Mental practice may be effective, he explains, "because it produces very small muscle contractions similar to those involved in physical practice. Images produced in the mind transmit electrical impulses to our muscles and tendons for the performance of an athletic exercise or event." (Helgoe 2006: 89)

This author advocates "blueprinting everything" because "it can create a similar mental path for any experience to follow, whether that's walking into a job interview, making a presentation or asking for an attractive person's phone number." She says, "because your subconscious doesn't distinguish between actual events and imagined ones, it triggers positive feelings and reactions generated by the 'past success'—the one you visualized—instead of panic or paralysis (*Ibid.* 88).

Reading this kind of thing, we might imagine that all we have to do to become professional dancers, skaters, athletes, or gymnasts is to "vision" it, although she does say that "visualizing your ideal reality may bring up a lot of enthusiasm and excitement, and it may also bring up some fears. And what if you have trouble conjuring up a detailed, positive visualization? Try focusing on feelings of accomplishment instead, or imagine receiving praise and congratulations on your success, suggests Vicky Thompson, author of *Life-Changing Affirmations: A 30-day Plan for Spiritual Transformation* (Red Wheel, 2005). . . . Transforming a vision from mind to matter is one of the fastest ways to make it feel, and become, real. This is the essence of the 10-step visioning method created by Lucia Capacchione, author of *Visioning: Ten Steps to Designing the Life of Your Dreams* (Tarcher/Putnam, December 2000)" (Helgoe 2006: 89).

To conclude this brief excursion into different usages of imagery, I will recall something I said about Dr. Sweigard and ideokinesis on page 45 of this volume. To my certain knowledge, she felt the same about other areas of human endeavor: "spiritual, the mystical (or occult), the artistic, and the psychological." Apart from that, I would want to say that aspiring dance students who arrive in university courses or dance academies to study dancing with these kinds of ideas of imagery foremost put themselves in a position of great disadvantage, because the teacher is faced with the task of undoing the damage that has been done. Even more damaging is the notion of imagery advocated by Franklin (1996), who seems to want us to believe that any and all imagery used at any time in any dance class is a "good thing."

Back to Ideokinesis

"Thinking"—the ideas (images)—*without any muscular activity whatsoever* is the basis upon which Sweigard's ideokinetic practices rested. In her own words:

> **Ideo,** the idea, the sole stimulator in the process, is defined as a concept developed through empirical mental processes. The idea, the concept of movement, is the voluntary act and the sole voluntary component of all movement. Any further voluntary control only interferes with the process of movement and inhibits rather than promotes efficient performance. **Imagined movement** is best defined as an ideokinetic facilitator. (1974: 7—bold type is added)

However, as we learned in Chapter 4 (p. 54), Sweigard also said,

> 6. Imagined movement should be used both before and during voluntary movement. There is always a tendency to start the actual movement too soon, once it is understood, without taking time to concentrate first on the imagery being used. The sequence after the movement has been learned is (*a*) complete rest, (*b*) imagined action, (*c*) slow movement with continued concentration on imagery, (*d*) sudden release of effort, and (*e*) sensing the change, if any. (1974: 276)

The point is that dance teachers and students should become more conscious of the imagery they use,* attempting to discover *why* they use it, then amending it where necessary for the students' benefit and, above all, for the purpose of doing no harm.

Habits

When a movement (*any* movement, danced or otherwise) is learned so well that it can be forgotten, it has moved from conscious to *subcortical* control in the body's neuromuscular system. In other words, the move has become *habitual.* Among many concepts, this means that no accomplished dancer *has to think about the dancing* performed in rehearsals or onstage because the moves are habitual. As with most artists, dancers learn a technique (i.e., ballet, modern dance forms, Kathak, and such) so they can *forget* the technique and focus on the meanings, expression, and the aesthetic qualities of the danced performance, including characterization, motivation, and the like. Williams James puts the matter in terms of *simplifying* the actual moves so that conscious attention can be focused elsewhere:

> Dr. Carpenter's phrase that "our nervous system grows to the modes in which it has been exercised" expresses the philosophy of habit in a nutshell" (cited in James 1950[1890]: 112). It does so because 1. *habit simplifies the movements required to achieve a given result, makes them more accurate and diminishes fatigue.* Furthermore, 2. *habit diminishes the conscious attention with which our acts are performed.* (1950[1890]: 114)

James makes other pertinent comments about habit (see Appendix to Chapter 6, page 100).

Clearly, how and in what ways habits are developed is a far more complex subject than we can deal with here; however, Pamela Matt succeeds in describ-

*See Raiport, Grigori. 1988. *Red Gold: Peak Performance Techniques of the Russian and East German Olympic Victors.* Los Angeles: Tarcher.

ing part of the process for ideokinesis with unusual clarity, pointing out how Mabel Todd's original perception—"problems with vocalization in singing and public speaking were often the result of poor habits of posture"—convinced her to develop "a new concept of posture" (Matt 1996: 337) that was radically different from older notions. Matt includes Sweigard in the picture by saying,

> Dr. Lulu Sweigard, a student of Mabel Todd's, demonstrated in experimental studies that Todd's approach did produce measurable changes in body alignment in at least nine dimensions. Sweigard organized Todd's imagery and developed new ideas for visualization which specifically pertained to those nine dimensions of change (Sweigard 1974: 193–96). . . .
>
> Barbara Clark, another student of Mabel Todd's, was a pediatric nurse and nursery school teacher who initially related Todd's ideas and practices to motor development. . . . Later in her career, Clark also worked extensively with dancers and actors in New York and enlarged the literature of Ideokinetics immeasurably with the development of several manuals of image drawings and exercises (see Matt 1993), cited in Matt (1996: 338).

Dr. Sweigard herself put the matter this way:

> When a person imagines movement, putting forth no voluntary muscular effort to aid its execution, the coordinated action of muscles which produces the imagined movement will be patterned subcortically. Imagining the movement is *a thought process only; it involves no muscular effort by the subject,* because muscular effort interferes with the skeletal changes which the imagined movement is designed to produce . . . . To change skeletal alignment by imagined movement, therefore, one needs to know where and in what direction the spatial relationship of bones must be changed to attain a more efficient structure. Having established this, one needs only to concentrate on visualizing a movement that will promote these changes to occur in the various places in the body. It will take time to establish more efficient neuromuscular habits; they are attained only through repeated [mental] practice. (Sweigard 1974: 222)

It sounds simple, doesn't it? But it is anything but simple seen from the standpoint of a teacher facing an average ballet (or modern concert dance) class in studios, universities, and academies throughout the country. I would like to emphasize here that the training process for dancing has *never* been simple, partly because of the ways in which students are assigned to classes, but this subject is dealt with at the beginning of Chapter 7.

Here, it is important to note that under the subheading "Body Mechanics Defined," written in 1938, Dr. Sweigard said,

> Body mechanics is the work of muscles exerting power on the bones as levers, to produce motion and maintain the body's equilibrium in relation to both internal and external forces. . . . Body mechanics is automatic, taking place with little, if any direct control by the individual, and as a rule without his being aware of it. Body mechanics, then, is an expression of neuro-muscular habits. It is a conditioned neuro-muscular reflex action directed toward producing and maintaining equilibrium in the body as the play of forces affects it from within and without. For this reason, how one's body mechanics is functioning at any given time is dependent upon the neuro-muscular habits that have been built up in his body, and which in turn have been influenced by all of life's experience. (Sweigard 1938: 18)

To Sweigard, body mechanics and posture were meant to be studied together. They couldn't be separated, because an individual's posture was at all times throughout life an expression of *habits* of body mechanics.

> For this reason a desirable alignment of the body cannot be achieved at any moment by voluntarily shifting into, and "holding" parts of the body in what would seem to be, either to the individual or to the instructor, a good posture. *It is only by changing the habit patterns of the individual's neuro-muscular action, that is, by changing his body mechanics, that change in posture can be achieved.* (Sweigard 1938: 19)

The "forces" that Sweigard saw affecting body mechanics, hence posture, were fourfold—"mental, emotional, sensorial, and mechano-physical" (*Ibid.* 18)—and she was convinced that the body always functioned as a whole that is *not* equal to the sum total of the functioning of its parts. She said that "postural deviations cannot be considered as entities in themselves; and body mechanics must be treated as a whole, not in part" (*Ibid.* 19). While she was mainly interested, I think, in how the functioning of a human "mechanism" was like that of any other structure governed by the laws of physics, she was also aware that thinking and emotions influence muscle action, saying that "since the time of William James there has been an increasing amount of research to support this principle as a working basis for educational procedure" (*Ibid.* 21).

Appendix to Chapter 6
Walking Bundles of Habit

WILLIAM JAMES

Could the young but realize how soon they will become mere walking bundles of habits, they would give more heed to their conduct while in the plastic state. We are spinning our own fates, good or evil, and never to be undone. Every smallest stroke of virtue or of vice leaves its never so little scar. The drunken Rip Van Winkle, in Jefferson's play, excuses himself for every fresh dereliction by saying, 'I won't count this time!' Well! he may not count it, and a kind Heaven may not count it; but it is being counted none the less. Down among his nerve-cells and fibres the molecules are counting it, registering and storing it up to be used against him when the next temptation comes. Nothing we ever do is, in strict scientific literalness, wiped out. Of course, this has its good side as well as its bad one. As we become permanent drunkards by so many separate drinks, so we become saints in the moral and authorities and experts in the practical and scientific spheres, by so many separate acts and hours of work. Let no youth have any anxiety about the upshot of his education, whatever the line of it may be. If he keeps faithfully busy each hour of the working day, he may safely leave the final result to itself. He can with perfect certainty count on waking up some fine morning, to find himself one of the competent ones of his generation, in whatever pursuit he may have singled out. Silently, between all the details of his business, the power of judging in all that class of matter will have built itself up within him as a possession that will never pass away. Young people should know this truth in advance. The ignorance of it has probably engendered more discouragement and faint-heartedness in youths embarking on arduous careers than all other causes put together.

Source: *Psychology* [Vol.1] 1950[1890], p. 127.

Chapter 7
More about Teaching Dancing

There are features of any class in ballet and modern concert dancing (including ballroom dancing at a competitive level) that are, in my opinion, liabilities. These liabilities are inherent, first, in the composition of most, if not all, dance classes of any kind. Classes are usually divided, first, into age groups, then into beginning, intermediate, or advanced classes. These designators serve to divide different age groups into units according to an individual's familiarity with an idiom of dancing as well as their ability level, recognizing how many years they have studied, and such. But, in any given class in the United States (and, I assume, elsewhere), one finds all different bodily types (i.e., endomorphs, mesomorphs, and ectomorphs) and combinations of these. Teachers are left to deal with them as best they can.

Put in dancerly terms, especially with regard to ballet classes (although our discussion of bodily types could apply to any dance class), students will tend to be *arqué* or *jarreté* (bow-legged or knock-kneed, respectively).*

Briefly, the point Noverre made is that an *arqué* dancer should not focus on the same bodily images that a *jarreté* dancer does, because the anatomical structure of their bodies affects their proficiency with regard to different kinds of movements (see the Appendix to this chapter, page 113, for Noverre's explanation). The *arqué* dancer is usually good at jumps, turns, and beats. The *jarreté* dancer generally excels at *développés, adagio* moves, and anything requiring flexibility. Noverre had many pithy remarks to make about the study of anatomy and the teaching of dancing:

> I dare conclude that the study of anatomy will render clearer the precepts which he [a ballet master or mistress] will impart to the pupils he wishes to train: from that moment he will distinguish with ease the natural and habitual defects of

---

*These terms are taken from Noverre in his *Letters on Dancing and Ballets* (1930[1760]). It is also useful to see Sparger (1965: 78–83) on the problem.

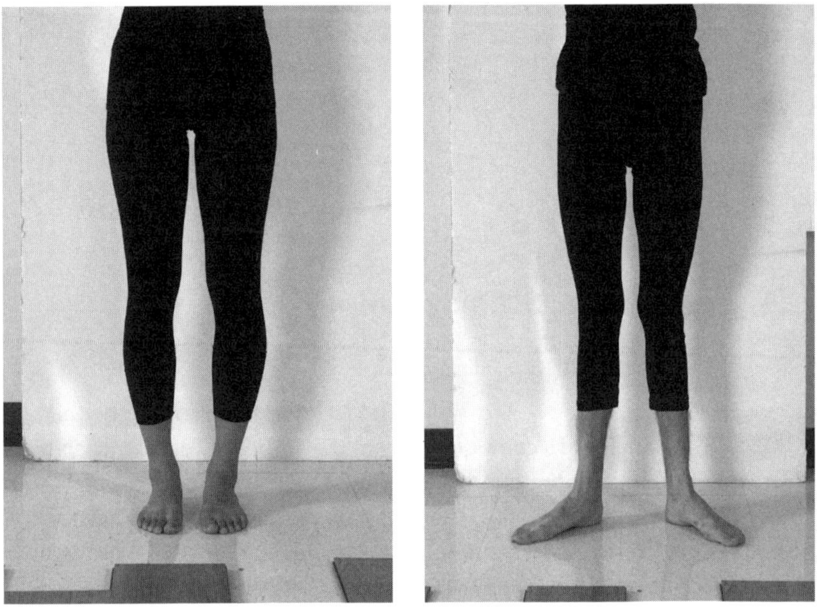

Images 16 and 17. Put in dancerly terms, especially with regard to ballet classes (although our discussion of bodily types could apply to any dance class), students will tend to be *arqué* or *jarreté* (bow-legged or knock-kneed, respectively). © Human Kinetics

> physique which so often impede a pupil's progress. Understanding the cause of the evil, he will easily remedy it; and, basing his lessons and counsels on a strict and careful examination, they cannot but be profitable. It is due to this lack of study of their pupils' physique, which varies as much as their physiognomies, that we owe that swarm of bad [ballet] dancers which undoubtedly would be less numerous, if care had been taken to place them in a suitable calling. (Noverre 1760[1930]: 35)

This paragraph provides a severely abbreviated view of Noverre's comments on teaching and the subject of bodily types, but, as I have included his comments in full in the Appendix to this chapter, p. 113), readers are welcome to see for themselves what he said.

I include Noverre's observations with regard to imagery in general as an example of what one should *never* encounter in a ballet or modern concert dance class: a teacher telling a student to 'feel' something or another in his or her body—for example, telling students how a high *développé* or some other move feels. Professional dancers who teach are often guilty of this. They seem to believe that if they tell students how a movement feels to them, the student will be better able to perform the move, whatever it is. Because they are successful dancers, they imagine that sharing their bodily sensations and/or kinesthetic images with students will be helpful; however, feeling (i.e., bodily sensation) images should, in my opinion, *not* be used in technique classes.

Imagine a *jarreté* type dancer telling an *arqué* dancer how something feels in his or her body: invariably, the *jarreté* teacher will not communicate with the *arqué* student—or the reverse. This is because the bones are aligned differently in *arqué* and *jarreté* bodies, thus the musculature in these types of dancer[s] is not the same. The tightness of tendonous attachments in legs and feet are different and much else; therefore, *one set of bodily sensations does not apply to the other.* Often, such 'sensation imagery' does even more subtle damage, because students who are the teacher's opposite bodily type will not feel what they are told they should feel, thus adding to what may be an already poor body image and a sense of failure. Nor is this the only problem.

I, and five of my students from Portland, Oregon, studied at Sadler's Wells School in England in the early fifties, where I first became acquainted with Celia Sparger's book and the necessity for different sets of images pertaining to diverse bodily conformations. I also want to point out (again) the relationship of anatomical study and the teaching of ballet:

> Why then should the modern teacher feel the need for the study of anatomy? For two reasons. Firstly, it is a sad but undeniable fact that when anything that is rare and precious becomes too easy of access, it is in danger of losing something in the process. Whereas at one time the few, the very few carefully selected, physically perfect and gifted children were chosen for initiation into the almost secret and almost sacred Schools of Ballet, and were taught by masters whose life was a dedication to the Art of Dancing and the Art of Teaching, handing on its secrets by demonstration, word of mouth, rare engravings and precious manuscripts; now instead, we have the printed word, the camera, the cinema and the open class. The barriers are down. Ballet has become the property of the masses. (Sparger 1982[1965]: 9–10)

Ballet having been "the property of the masses" for some time now means that anyone who wishes can attend ballet classes regardless of their physical condition, which may not be suitable for the study of ballet dancing at all. A student's condition (age, history of injuries, and such) may not be suitable for the study of professional modern concert dancing, either. This is to say that well over 50 percent of the people who attend ballet classes, Graham technique classes, or professionally based classes of any kind will never be (nor will they come close to being) professional dancers. This is generally true of classes taught in universities.

Many students enroll in university dance classes because they believe that ballet (or modern dance) training is more beneficial than gymnastics or sports. They may prefer danced moves over physical-education activities, or they believe that dancing will improve their physical health and strengthen their bodies. This may be accurate for some of the physically better-equipped students, but it is not the case for all aspiring students.

Some students take dance classes because they want to know more about the idiom of dancing so that they can increase their appreciation of professional performances. Some are there because they want physical exercise but do not care for competitive team sports. There are as many different reasons for studying ballet or modern concert dancing at a student level as there are people attending the classes. Somewhere along the line, the few who have some hope of actually doing professional dancing—of getting into a company, say, or getting a job in a musical or something of that nature—get separated from the many who do not. As Gelsey Kirkland points out, many fall by the wayside even in professional private studios or company-connected schools:

> Balanchine and his teachers unwittingly encourage young dancers to self-destruct, rationalized as part of the sacrifice that must be made to the art. The speed and shortcuts that he built into the training process called for physical cheating in which the dancer distorted the body to deliver the position or step that Balanchine demanded. The risk of injury was ignored. I watched many of my friends become casualties and fall by the wayside. (Kirkland 1987: 28–29)

Whoever can pay fees for classes is allowed to attend. A dance teacher's living is made not out of the few who become professionals, but the many who do not. Where do teachers of such classes stand in relation to principles of good training, no injuries and such, given that most of them are interested in serving the needs of as many students as they can? I think Sparger sums up the differences between many ballet teachers' approach to teaching (using dance imagery) and teaching that uses anatomical imagery when she says,

> Curiously enough, anatomy seems to suggest to most people the study of muscles. Actually for the teacher of ballet this is of very little importance, for the simple reason that ballet has its own technique of definite and exact movements. If these movements are performed correctly the correct muscles will work. (Sparger 1982[1965]: 10)

But what of students in the class who, because of anatomical difficulties (i.e., pronated ankles, mild scoliosis, previous injuries, or other types of faulty postural deformations), may not be able to perform the movements correctly? This is where I think that the system of body techniques and awareness I devised came into play in classes—not for those unfortunates who had such severe distortions that they couldn't dance very well, but for those with milder deficiencies.

Those procedures (Chapter 1, pages 6–10) were there so that I might assist in whatever way I could to improve the lot of imperfect bodies as well as teach students to protect themselves from misguided imagery in classes. I did not, of course, teach perfect bodies. *No one in any field of human movement teaches perfect bodies.* Even students who are taken into the Royal Ballet School (formerly Sadler's Wells) aren't perfect. But here, one runs up against another problem of teaching in the United States.

Comparatively few schools on this side of the Atlantic have companies attached so that company classes are able to be taught, thus separating those who have somehow reached an advanced or professional level. New York City Ballet and the American Ballet School had such schools, but unfortunately, they were headed by George Balanchine and Mikhail Baryshnikov respectively, neither of whom had any idea, really, about *teaching* dancing, except for the received sets of images, attitudes, and so forth that they themselves had been taught. Before going on, it is necessary to say that George Balanchine died in 1983. Baryshnikov hasn't been connected with the American Ballet School since 1990, therefore the thrust of my comments pertains to the consequences of these artists' teaching, *not* to the present historical contexts of the schools.

There is no doubt that Balanchine was a brilliant choreographer, but all he really wanted was a group of mindless dancers who would perform his choreography as he wished it to be performed—and only that. He didn't want dancers to think. He was a complete martinet in classes. In a revealing chapter on her early training, Gelsey Kirkland observes:

> I have a vivid memory of one of the teachers stopping me in the middle of class to demand that I turn out my feet. There was no regard for the knees or hips, which in my case were distorted to the breaking point. The teacher refused to continue the class until I complied with her wishes. . . . During my third year of training, I was introduced to toe shoes, the standard footwear for all ballerinas since the mid-nineteenth century. . . . Misguided and caught between excessive demands for turn-out and *pointe,* my feet had already begun to deform. At the age of eleven, I came down with a severe case of bunions. Many of the teachers had the same malady, caused from years of strain placed on the foot. It was said that Balanchine cherished the aberration of line induced by bunions, that they contributed to the impression of winged feet. (Kirkland 1986: 34–35)

Notice the image of turning out the *feet*—contrary to the principle of directing images to the appropriate part of the body that actually performs the task.

On the other hand, Barishnikov never really cared whether anyone else danced well or not unless he was partnering them. To me, he was an example of a great dancer who should never have tried to teach, except, perhaps to teach roles from ballets—*never* technique.

With regard to imagery and the teaching of dance classes: the reason I developed the locational techniques and centralizing light-beam images was because these were the only ways I could see to work out of some of the physical dilemmas that are (unfortunately) endemic in the dance world. My use of locational and centralizing images based upon Sweigardian principles was aimed toward teaching each individual how to protect him- or herself in dance classes.

Since there is nothing that any single dance teacher can do about altering the situation I've outlined above, the best advice I can give American teachers of ballet and modern concert dance classes is, first of all, DO NO HARM. But doing no harm in a ballet or modern concert dance class is a thorny, demanding task, partly because many movements required for ballet dancing or modern concert dancing *are not beneficial to the body* from an ideokinetic point of view. It is interesting that Matt describes this fact in her discussion about Clark's career as a teacher of kinesthetic awareness in New York:

> Joanne was also studying with José Limon. As she progressed into Limon's advanced classes, her body began to rebel. A light case of polio in childhood had caused some muscle weakness and the strenuous dance movement, "was not making me better, it was making me worse." The leg that was affected by the polio began to collapse mysteriously in leaps across the floor and her back pain was constant and often excruciating. Out of concern for a promising student, Limon's wife, Pauline Lawrence took Joanne to Sweigard and then to Barbara [Clark]. (Matt 1993: 47)

The upshot is this: After I had studied with Sweigard for a year, I found myself in a position of having to choose whether I remained a teacher of dancing or became a teacher of ideokinesis, more interested in teaching bodily principles and kinesthetic awareness to everyone, whether they were dancers or not. I chose to remain a teacher of dancing.

Different though their approaches to ideokinesis were (see Matt 1993: 44–45), Clark and Sweigard were not dancers, nor did either of them understand Limon's technique, Graham's technique, or the ballet as performing art disciplines all that well. I would want to say, however, that a person with the kind of kinesthetic awareness that Clark advocated might have developed a technique of dancing that would be different from anything known in the history of dance forms in the United States. In spite of this, with regard to older, highly sophisticated idioms of dancing, I am prepared to say that ideokinesis is therapeutic, and of immense *remedial* value, but it does not have any appreciable *direct* effect on the movements that are required to be performed *if* the ballet or the various techniques of modern concert dancing are to maintain their integrity.

Sweigard corrected the underlying postural patterns and habitual moves of individuals who, when subjected to the demands of highly complicated, difficult dance techniques, experienced strain (often breakdowns), but she did not deal with the dance idiom directly. Clark's services were used "to relieve the discomfort of injuries or as a means of recovery from a long season of performances. Others [e.g., Erick Hawkins] became committed to her approach" (Matt 1993: 46).

There can be no doubt that Sweigard's and Clark's imagery improves the dancer's awareness of his or her own body (and the bodies of others), but in the last analysis, their imagery does not produce ballet dancing or modern concert dancing, any more than it would produce Olympic gymnasts, champion figure-skaters, or competitive ballroom dancers. Recreational forms of dancing (that is, ballroom dancing, folk dancing, and such) are more amenable to those whose main interest is in developing kinesthetic awareness together with the all-pervading sense of well-being that kinesthetic awareness supports in everyday life.

In other words, with regard to ballet and modern concert dancing, ideokinesis plays an important rehabilitative and enabling role in some cases, because it can alter or remove some of the habitual neuromuscular problems that are inherent in any dancer's posture. It therefore enables some dancers to be more proficient in their art because, through Sweigard's ideokinesis, they can get rid of habitual tensions and imbalances that inhibit the best potential performance in their chosen idiom of dancing and in everyday movements as well.

This is *not* to say that dancers do not have (or develop) high degrees of kinesthetic awareness, whether they study ideokinesis or not, because they do. It *is* to say that the demands of the bodily discipline of the form of dance that is chosen must supersede the requirements of kinesthetic awareness as it was taught by Todd, Sweigard, Clark, Lawrence, Litz, Bernard, and others. As far as I am aware, Erick Hawkins's usage of Clark's imagery in his modern dance technique is available in Celichowska's book (2000). To some extent, she explains how he achieved the effects he did, but there is very little about how he applied ideokinesis to (or in) his classes.

Even so, the influence of Sweigard's and Clark's work over dancers was (and still is) in the dance world on a fairly large scale. According to Matt (and I believe she is accurate), it extended to "the unrest in the world of modern dance":

> In the late sixties and early seventies, many of [Clark's] students were of the "post-modern dance" or "new dance" persuasion [Mary Fulkerson, David Woodberry, and Nancy Topf were among these]. They were disenchanted with the traditional approaches to technique which damaged bodies and accepted physical suffering as the way to improve. The social and political traditions of dance were also questioned by these young people. Traditional dance protocol placed choreographers and teachers in positions of unquestioned authority. Students, attempting to progress through this system, often experienced the loss of personal identity in the attempt to please their elders. Rejecting old imitative dance technique conventions and paternalistic hierarchies in companies and schools, students began to train exclusively through improvisation and to express their joy in dancing through loosely structured contact jams. (Matt 1993: 72)

Of course, no one in their right mind really believes that "damaged bodies and physical suffering" result in (or cause) improvement in dance classes. Unfortunately, many dancers *do* wear their injuries and pain as if they were badges of honor. "No pain—no gain," is, I believe, the common euphemism among dancers, gymnasts, and others.

It is also difficult to accept the view that all of someone's woes in the dance world are the result of unquestioned authority. Surely the loss of personal identity is based on more than pleasing one's elders. While it is true that many acceptable imitative dance technique(s) are naive and immature, students can be encouraged to resort to other measures than the rejection of *all* disciplined dance techniques for the kind of movement chaos that many postmodern dance concerts and performances display.

Naomi Jackson (2000) offers the only lucid assessment of postmodernism in the dance world that I have so far encountered:

> Today there are two opposing poles of postmodernism with contrary ethics: one advances the idea of a pluralist, egalitarian America that is both cultured and tolerant; the other, an America where an identity politics of resentment thrives. The first view fosters an inclusive, cosmopolitan America that integrates difference into itself: the second acclaims difference to the exclusion of all else. The first takes the newer insights of cultural theory and poststructuralism regarding the mechanisms of cultural production and dissemination of power as a means of creating a world of greater equality and compassion. The second revels in the gaps, holes, fissures, and fragmentation of so-called cultural texts

> to decry the reality of individuals, meaning, truth and even history. Given these differences, it is clear that the Y acted as a forerunner of the first mode of postmodernism, reflecting as it did a humanist tradition in Jewish thought and ethics that strove to create a framework for equity and for individual and social betterment, rather than a space of separation, nihilism, skepticism, or lack of meaning. (Jackson 2000: 210–11)

Elsewhere in her admirable book, Jackson says,

> Instead, the choreographers were, as the dance historian Sally Banes has written, engaged in finding "new ways to foreground the medium of dance rather than its meaning. In her introduction to *Terpsichore in Sneakers: Post-modern Dance,* Banes traces the connection of the Judson dancers to a larger trend away from meaning in art. ". . . The younger generation of choreographers showed in their dances that they departed not only from classical modern dance with its myths, heroes, and psychological metaphors, but also from the elegance of ballet." Choreographers like Yvonne Rainer, Steve Paxton, and David Gordon . . . were engaged in saying no to everything that came before, claiming that you did not need to have virtuosic technique or use dramatic narratives in order to be a dancer or choreographer. (Jackson 2000: 216)

Fortunately, just as the tradition of ballet continued through the rebellions of Isadora Duncan and a long, distinguished line of modern concert dancers over the years, it will, I trust, continue through the postmodern phase as well, along with the traditions of modern dance techniques well established by Humphrey-Weidman, Graham, Limón, Tamiris, and many others. This does not mean that I advocate teaching styles akin to those of Balanchine or Baryshnikov, nor does it mean that I defend *any* physically based discipline (in sports or dancing) whose teachers consistently injure a majority of its participants. However, I do not, because of that, want to see the arts of ballet dancing or modern concert dancing eliminated from the world of dance performance and the performing arts.

What I *am* saying is that ballet dancing and modern concert dancing need not be dominated by paternalistic—or maternalistic—hierarchies and suffering. These forms of dancing need not be accompanied by loss of identity, anorexia, tendonitis, bunions, constant pain, and such. These are aberrations of individual teaching styles and attitudes—*not the structures of the dance forms themselves.*

Poor, unknowledgeable, and/or mediocre teaching is deplorable in any field, including the dance. In fact, the subject of teaching any physical activity is extremely difficult, and I wonder if there are articles in physical education or somatics literature about the deleterious effects of, say, Béla Károlyi's teaching of Olympic gymnasts? To me, his results for young gymnasts are much the same as Balanchine's were with regard to ballet dancers: broken, injured, or distorted bodies.

Any enlightened system of teaching movement, whether ballet, modern concert dancing, championship ice-skating, or gymnastics, requires a high degree of anatomical knowledge and kinesthetic awareness, but we should be clear about this: the teaching of kinesthetic awareness with the aim of improving one's life and all the moves one makes in everyday life is not the same as teaching a dance idiom that ultimately produces performing artists of surpassing excellence.

Moreover, not all teaching all over the world is the same. The teaching of ballet in England is different from ballet teaching in Canada and the United States because of a relatively higher degree of standardization of teaching and exams for all ages and grades of student, *if* the teacher belongs to the RAD (Royal Academy of Dancing), the Vaganova or Cecchetti societies, or to an organization such as the Royal Ballet (formerly Sadler's Wells). Such requirements in all three countries are, however, optional, and the standardization of teaching dancing has many adversaries throughout the world.

In Canada, as in the United States, teachers can open studios and maintain them if they do not break any *civil* laws—none of which have anything to do with teaching dancing without injury. In California (and perhaps some other states), one is required to have a "dance hall" license because there is no differentiation at this public level between "dance halls" and dance schools, studios, or academies. Schools of ballet and/or other forms of dancing can be opened in Canada and the United States on a basis of personal whims, emotions, local acclaim, and the like. There are no general criteria for opening schools based upon informed or enlightened *teaching*. During the years I taught, dance teachers did not have to be accredited with reference to the bodily techniques they advocated, whatever those techniques might have been.

On the whole, it may be the case that Canadians are more aware of dance teaching organizations than Americans. This may be because of a smaller population, and the genuine educational efforts made by major Canadian dance companies and their respective schools and auditions in Canada. There were strong views on the values of the teaching organizations in both countries. Not everyone would welcome institutionalization or any kind of accreditation. Years ago, I had no choice regarding certification with regard to anatomical knowledge, but that situation has changed since 1983:

> The Aerobics and Fitness Association of America (AFAA) was one of the first national organizations to provide dance exercise certifications on a large scale—the Primary Aerobic Instructor Certification. . . . Another new and very suc-

> cessful organization is the International Dance Exercise Association (IDEA), which was established in 1982. . . . The IDEA Foundation offered its first Aerobic Dance Exercise Instructor Certification Exam in 1986 and by the end of the first year approximately 2,000 instructors were IDEA certified. . . . Their current membership encompasses 23,000 fitness professionals in 88 countries. (Clippinger 1997: 44)

Clippenger goes on to say that in spite of the long history of dance forms, "it is surprising that we have not developed parallel standards and certification or licensing for the dance instructor. Although some of the goals of dance and fitness differ in that dance also has artistic goals, it is still the same human body that is the tool for expression, no matter what the mode of movement. The goal is not to censor artistic expression but rather to wed scientific principles with the many differing forms and expressions of danc[ing] so that dance training and performance can be enhanced" (Clippenger 1997: 45).*

The issues that have briefly been raised here are 1. licenses to teach (or lack of same); 2. the teaching of dancers so that they can protect themselves from unskilled, unknowledgeable, and/or emotionally unscrupulous teachers, and 3. (in the world of professional dancing) the need for a separation of teaching roles from ballet and other repertoires and the teaching of daily technique classes.

The fundamental topic I have raised, however, has to do with teaching itself. *Teaching* dancing requires knowledge that dancers themselves need not have, and the teaching of ballet and any form of modern concert dancing is both science and art, but the teaching of dancing has rarely, to my knowledge, been considered to be a science *or* an art in the United States, except, perhaps, by a few. For instance, if someone has danced professionally, then it is automatically assumed that he or she can teach. Such persons can teach, of course, *if* all that is required is the knowledge of steps, *enchaînements*, and such. But should this be the only requirement?

Balanchine's regrettable teaching practices caused me to stop attending New York City Ballet's performances years ago. I no longer watch American Olympic gymnastic displays, either, as Nadia Comăneci's coach, Béla Károlyi, is (or was) Balanchine's counterpart in American gymnastics.

The spectacle of driven, neurotic, bandaged youngsters on a vault or balance beam is no more acceptable than visions of anorexic, obsessed dancers with bunions on a concert stage. Improperly taught, ballet dancing and mod-

---

*A word of caution here: excellent though the goals of the association are, there is a problem in terms of some of the exercises students are asked to do. There needs to be as much *focus on correct alignment and proper execution of the movements* as there is on heart rates.

ern concert dancing (unlike many other arts) have the potential of harming their executants. The teacher's role as an 'instrument maker' can wreak havoc among students. Dancing has its dark side, so to speak, and until that is addressed, I doubt that the teaching of dancing overall will improve.

So, What Are We to Make of It All?

With regard to the subject of bodily imagery, I have deliberately introduced the liabilities of teaching dance classes in the United States, because in a democratic society such as ours, it is hypothetically possible for anyone to become a modern concert dancer or a ballet dancer if they want to—just as it is possible for any person who so desires to become president.

The realities of both situations are, of course, quite different. While it is true that anyone can attend dancing classes if they can afford to pay for them, the moment an aspiring student steps into a class, his or her individual bodily posture (plus age, history of injuries, and such) becomes of paramount importance whether the individual student (or the teacher) is aware of the problems and potentialities or not; however, the fate of any given student depends very largely upon an individual teacher's ability, knowledge, understanding, and expertise.

The world of the dance classroom: the "dance technique class" is all-important, yet this classroom has been assigned minimal importance in the literature and in reality. Of the thousands of students who enter dance classes every year, only a few will persevere to become professionally skilled dancers. Given the freedom to express unwelcome truth, I would say that the American public doesn't give a tinker's curse about the intricacies of the dance classroom.

Student dancers who don't make it in the profession are *invisible*, just as their problems, in general, are both invisible and unexamined. As Gelsey Kirkland's book so accurately illustrates, even the physical problems of ballerinas—celebrities who have 'made it'—are invisible until someone who has lived through them and survived draws attention to them.

Appendix to Chapter 7
Letter XI

JEAN-GEORGES NOVERRE

[NB: C.W. Beaumont's spelling and punctuation for the translation have been retained.]

It is rare, Sir, not to say impossible, to find men perfectly proportioned, and for this reason it is quite common to meet with a crowd of ill-proportioned dancers, in whom one only too often sees defects of physique which all the resources of art can scarcely conceal. Is this due to a fatality of human nature by which we always withdraw from that which is suited to us, and so often pursue a career in which we can neither advance nor succeed? It is this blindness, this ignorance concerning ourselves, that produces such a multitude of bad poets, of mediocre painters, of tedious players, of noisy musicians, of detestable dancers and mummers, in short, Sir, of impossible people in every art. These same men put in suitable posts might have been useful, but, away from their proper occupation and station, their real talents are wasted and exchanged for ones which vie with each other in absurdity.

When one wishes to enter into the dancing profession, the first consideration to be taken is that of physique, provided that the postulant has attained an age at which he should be capable of reflection. If his natural defects be irremediable, he must renounce at once, and completely, the thought conceived of the advantage of competing in the pleasures of others. Should these defects be capable of remedy by care, constant exercise, and by the counsel and advice of a well-informed and enlightened teacher, then it is imperative not to neglect a single effort which can remedy the imperfections over which he desires to triumph. He must look forward to the time when his body has attained its greatest strength and stability, that is to say, its full growth; and when his desire to conquer his defects has been strengthened by so long and inveterate a habit that it can no longer be destroyed.

Unfortunately there are few dancers capable of this stringent self-examination. Some, blinded by self-esteem, imagine themselves to be faultless; others shut their

---

Source: Jean-Georges Noverre's *Letters on Dancing and Ballets*, 1930[1760]: 109–16.

eyes to those defects which the most cursory examination would reveal to them. Again, if they disregard what every man of intelligence has the right to reproach them with, it is not astonishing that they fail to achieve their aim. The disproportion of the limbs ceaselessly opposes the play of the muscles and the harmony which should make a perfect whole. There is no more union in the steps; no more softness in the movements: no elegance in the attitudes and oppositions: no proportion in the *déploiements* and consequently no firmness and no equilibrium. To that point, Sir, has been brought the execution of the dancers who are blind to their physique and who fear to examine themselves when they study and execute their exercises. We can, without offending them, give them their just due and call them bad dancers.

Probably, were good masters more common, good pupils would not be so scarce; but the masters who are fit to teach do not give lessons, and those who ought to take them are eager to give lessons to others. What shall we say of the careless uniformity of their teaching? One will cry, "There is only one correct method of teaching." I agree, but is there only one way of pointing this out and passing it on to pupils one likes, and is it not essential to lead them to the same end but by different roads? I acknowledge that to succeed in it one must have a particular sagacity, for, without reflection and study, it is impossible to apply principles according to the different types of conformation and the various degrees of aptitude; one cannot see at a glance what is expedient for one, what is not suitable for another, and lastly one can never vary the lessons in proportion to the different varieties of which nature or habit, the latter often more rebellious than nature even, offer and present us.

It is then essentially the master's right to place each pupil in the class suitable to him. For this purpose it is not sufficient to possess only the most exact knowledge of the art, it is also necessary to guard against that vain pride which inclines each one to think that his own method of execution is the best and the only one which can please; because a master who always regards himself as a model of perfection and devotes himself only to making of his pupils a copy of which he is the good or bad original, will not succeed in making even tolerable ones, except he encounter those who are gifted with the same disposition as he, and who have the same height, the same conformation, the same intelligence, and lastly, the same aptitude.

Among faults of physical construction, I generally notice two principal ones; one is to be *jarreté*,* the other is to be *arqué*.† These two defects of conformation are almost universal and differ only in their degree; we see very few dancers free from one or the other.

A dancer is said to be *jarreté* when his hips are narrow and turned inwards, his thighs are close together, his knees are large and so close to each other that they touch and almost adhere, while his feet are quite apart from each other so that the space between the knees and the feet almost resembles the figure of a triangle; further, I remark an enormous bulge in the inner surface of his ankles, while his instep is

---

*Literally, *gartered* (or, as we would say, "knock-kneed").

†Literally, *arched* (or, as we would say, "bow-legged").

higher than usual, and his tendon Achilles is not only slender and puny, but very far from the joint.

A dancer who is *arqué* is one in whom the contrary defect is noticed. This defect continues from the hip to the foot, for these parts describe a line which gives the figure some resemblance to a bow; in fact the hips are broad, the thighs and knees well apart, so that the light that should naturally be seen between some of the portions of the lower limbs when closed, is seen all the way down and appears wider than it should be. Persons so constructed have in addition a long and flat foot, the outside of the ankle projecting, and the tendon Achilles large and near the joint.

These two diametrically opposed defects prove with greater force than any rhetoric that the lessons suitable for the first would be harmful to the second, and that the studies of two dancers as different in height as in physical structure, cannot be the same. The one who is *jarreté* must continually devote himself to separating those parts pressed too closely together. The first method of succeeding in this is to turn the thighs well outwards and to move them in this direction, while profiting by the liberty of the rotary movement of the femur in the cotyloid cavity of the hip-bones. Aided by this exercise, the knees will follow the same direction and return, as it were, to their place. The knee-cap, which seems intended to limit the flexion of the knee too far behind the joint, will fall perpendicularly over the point of the foot, and the thigh and leg, being in the same line, will thus describe a perpendicular which will assure the firmness and stability of the body.

The second remedy to employ is to maintain a continual movement in the knee joints, so that they appear extremely stretched without actually being so. That, sir, is a work of time and habit; when this habit is strongly contracted, it is impossible for the joint to retake its natural and vicious position without efforts which cause in those parts an insupportable numbness and pain. I have known dancers who have discovered the art of concealing this defect to such a degree that it would never be noticed if the straight *entrechat* and very strong *temps* had not betrayed them. This is the reason: the muscular contraction resulting from the efforts of jumping, tightens the joint and forces each part to go back into its place and return to its natural form. The knees, thus forced are carried inwards, they reassume their normal size, which offers an obstacle to the beats of the *entrechat;* the more these parts come together, the more those which are beneath them are separated; the legs, being unable either to beat or to cross, remain as if motionless at the moment of the action of the knees which roll disagreeably one on the other, and the *entrechat* being neither *coupé,* nor *battu,* nor *croisé* by the step, cannot have the speed and brilliancy which are its merit. Nothing is more difficult, in my opinion, than to conceal these defects, above all in vigorous movements where the whole body is shaken and sustains violent and reiterated jerks, giving itself to contrary movements and to continual and varied efforts. If art then can triumph over nature, of what praise is not the dancer worthy?

He who is thus constructed should renounce *entrechats, cabrioles* and all difficult and complicated steps, with all the more reason that his execution will certainly be feeble, because his hips being narrow, or, to employ the language of anatomy, the pelvic bones being less wide, they allow less play to the muscles which are attached to them and on which the movements of the body partly depend; movements and flexions are much more easy when these same bones are much wider, because then the muscles end or start from a point further distant from the centre of gravity. However that may be, the *danse noble* and *terre à terre* is the only one suitable to such dancers. Besides, Sir, what dancers who are *jarretés* lose in strength, they seem to gain in grace. I have noticed that they were soft and brilliant in the most simple things, easy in difficulties which did not require efforts beyond their powers, and that their beats are always graceful because they make use of, and profit by, the bones and tendons which move the instep. These are the qualities which compensate them for the strength they lack; and, in matters of dancing, I always prefer skill to strength.

Those who are *arqués* have only to apply themselves to bringing together those parts which are too separated, to diminishing the space which is found principally between the knees. They have no less need than the others of the exercise which turns out the thighs, and it is less easy for them to conceal their defects. In general, they are strong and vigorous; consequently they have less suppleness in the muscles and their joints move less easily. It can be understood that if this defect in physique arises from the deformity of the bones, all labour would be useless and the efforts of art powerless.

I have said that dancers who are *jarretés* should maintain a little flexion in their execution, the former from the opposite cause should be fully stretched and cross their *temps* more narrowly so that the union of the parts can diminish the light or interval which naturally separates them. They are sinewy, lively and brilliant in things which demand strength rather than skill; sinewy and light, considering the direction of their groups of muscles, and considering the consistency and resistance of their articular ligaments; lively, because they cross rather low than high, and, having for this reason less distance to go to beat the *temps,* they pass them with more speed; brilliant, because of the light between the parts which cross and uncross. This light, Sir, is exactly the *chiaroscuro* of the dance, because, if the beats of the *entrechat* be neither cut nor beaten, but on the contrary, rubbed or rolled one on the other, there will not be the light which gives value to shadows, and the legs, too close together, will offer only an indistinct and ineffectual mass. They have little skill because they rely too much on their strength, and this same strength is opposed to their suppleness and ease. If their strength leaves them for an instant, they are clumsy; they ignore the art of concealing their state by using simple *temps* which, requiring little effort, always give them the time to regain their strength; moreover, they have very little elasticity and rarely strike with the toe.

I believe I have discovered the true reason for this when I consider the long and flat shape of their feet. I compare this part to a lever of the second order, that is to say, a lever where the weight is between the fulcrum and the power, so that the fulcrum and the power are at its extremities. Hence the fixed point or fulcrum corresponds to the extremity of the foot, the resistance or weight of the body is carried on the instep, and the power which raises and sustains this weight is applied to the heel by means of the tendon Achilles. Now as the lever is stronger in a foot that is long and flat, the weight of the body is further distant from the fulcrum and nearer the power; hence the weight of the body must be greater and the strength of the tendon Achilles diminish in equal proportion. I have said then that this weight not being in as exact a proportion in dancers that are *arqués* as it is in those who are *jarretés,* who have usually a high and strong instep, the former have necessarily less ease in raising themselves to the extremity of the toes.

I have observed again, Sir, that defects encountered from the hips to the feet make themselves felt from the shoulder to the hand, more often the shoulder follows the conformation of the hips; the elbow that of the knee; the wrist that of the foot. A little attention will convince you of this truth, and you will observe that, in general, defects of physique, resulting from the vicious arrangement of certain joints, are common to all. This principle accepted, the teacher should suggest different movements of the arms to different pupils. This care is most important; short arms require only movements proportionate to their length, long arms cannot lose their extent except by the roundness given them. Art consists in taking advantage of these imperfections, and I know dancers who, by means of the *effacements* of the body, skilfully conceal the length of their arms; they made a portion of them recede into the shadow.

I said that dancers who are *jarretés* are feeble, they are thin and slender; dancers who are *arqués* are more vigorous, they are big and sinewy. It is a common failing to think that a man who is big and thick-set must be heavy; this principle is true as regards the actual weight of the body, but it is false in what concerns dancing, because lightness is born solely of the strength of the muscles. Every man who is only feebly aided by them will always 'fall' heavily. The reason is simple, the weak parts being unable to resist the stronger parts at the moment of the fall (that is to say, the weight of the body which acquires, according to the height from which it falls, a new degree of weight), give way and bend. It is in this moment of relaxation and flexion that the noise of the fall is heard, a noise which diminishes considerably and which can even pass unnoticed when the body can maintain itself in an exactly perpendicular line, and when the muscles and tendons have vigour enough to oppose the falling weight and to resist vigorously the shock which might cause them to give way.

Nature has not spared the fair sex from the imperfections which I have mentioned to you, but artifice and the fashion of petticoats have happily come to the aid of our *danseuses.* The panier conceals a multitude of defects, and the curious glance of the critics cannot rise high enough to pass judgment. Most *danseuses* dance with the knees separated, as if they were naturally *arqués.* Thanks to this bad custom and to petticoats, they appear more brilliant than the men, because, as I have said, beating only with the lower part of the leg, they pass their *temps* quicker than the latter, who, concealing nothing from the spectator, are obliged to beat theirs with stretched muscles and to make them come chiefly from the hips. You understand that it takes more time to move a whole than a part. As to the brilliancy which *danseuses* evince, vivacity contributes to it, but less, however, than the petticoats, which, concealing their defects of physique, fix the looks of the spectators more attentively. All the vigour of the *battements* being, as it were, brought to one point, they appear livelier and more brilliant; the eye can take it all in at a glance, which is the more concentrated the smaller the space it has to traverse.

Besides, Sir, a pretty face, beautiful eyes, an elegant form and voluptuous arms, are the inevitable rocks on which criticism founders, and powerful claims to the indulgence of the spectator, whose imagination substitutes for the pleasure which he has not received, that pleasure which he might possess of the stage.

I am, &c.

References Cited

Ahsen, Ahkton
1984. Imagery, Drama and Transformation. *Journal of Mental Imagery* 8(1): 53–78.

Banes, Sally
1987. *Terpsichore in Sneakers: Post-Modern Dance.* Hanover, NH: University Press of New England (Wesleyan).

Blom, Lynne Anne, and L. Tarin Chaplin
1985. *The Intimate Act of Choreography.* Pittsburgh: University of Pittsburgh Press.

Bonpensiere, Luigi
1953. *New Pathways to Piano Technique: A Study of the Relations Between Mind and Body with Special Reference to Piano Playing.* New York: Philosophical Library.

Celichowska, Renata
2000. *The Erick Hawkins Modern Dance Technique.* Hightstown, NJ: Princeton Book Company.

Clippenger, Karen S.
1997. Standards and Certification for the Dance Instructor. *Journal of Dance Medicine and Science* [JDMS] 1(2): 44–45.

Dowd, Irene
1981. *Taking Root to Fly: Seven Articles on Functional Anatomy.* New York: Self-published.

Farnell, Brenda
1996. Metaphors We Move By. *Visual Anthropology* 8(2–4): 311–36.
1999. It Goes without Saying—But Not Always. In *Dance in the Field: Theory, Methods, and Issues in Dance Ethnography* (Ed. Theresa Buckland). New York: St. Martin's Press, pp. 145–60.
2003. Kinesthetic Sense and Dynamically Embodied Action. *JASHM* 12(4): 132–44.

Fitt, Sally Seavey (Ed.)
1996. *Dance Kinesiology.* 2nd edition. New York: Schirmer Books.

Fleshman, Bob
1984. Mime and Movement Training with Mental Imagery. *Journal of Mental Imagery* 8(1): 91–98.

Foucault, Michel
1988. Technologies of the Self. In *Technologies of the Self: A Seminar with Michel Foucault* (Ed. Luther Martin). Amherst: University of Massachusetts Press, pp. 16–49.

Franklin, Eric
1996. *Dynamic Alignment through Imagery.* Champaign, IL: Human Kinetics.
H'Doubler, Margaret
1962. *Dance: A Creative Art Experience.* Madison: University of Wisconsin Press.
Hanrahan, Christine, and John Salmela
1990. Dance Images—Do They Really Work or Are We Just Imagining Things? *Journal of Physical Education, Health, Recreation and Dance* [JOPEHRD] 61(2): 18–21.
Hart-Johnson, Diana
1995 (with Rajika Puri). Thinking with Movement: Improvising versus Composing. In *Human Action Signs in Cultural Context: The Visible and the Invisible in Movement and Dance* (Ed. Brenda Farnell). Metuchen, NJ: Scarecrow Press, pp. 158–86.
1997. A Graham Technique Class. *JASHM* 9(4): 193–214.
Havas, Kató
1964. *The Twelve Lesson Course in a New Approach to Violin Playing.* London: Bosworth.
Helgoe, Courtney
2006. See It, Believe It. *Experience Life,* January/February.
Hill, Martha
1961. Introduction to The Dancer's Posture, by Lulu E. Sweigard. *Impulse: The Annual of Contemporary Dance.* San Francisco, California.
Jackson, Naomi M.
2000. *Converging Movements: Modern Dance and Jewish Culture at the 92nd Street Y.* Hanover, NH: University Press of New England (Wesleyan).
James, William
1912. *Radical Empiricism.* London: Longman and Green.
1950[1890]. *Psychology* [Vol.1]. London: Macmillan.
Kirkland, Gelsey, with Greg Lawrence
1986. *Dancing on My Grave.* New York: Jove Books.
Lakoff, George, and Mark Johnson
1987. *Metaphors We Live By.* Chicago: University of Chicago Press.
Lalli, Gina
2004. A North Indian Classical Dance Form: Lucknow Kathak. *Visual Anthropology* 17(1): 19–44.
Lee, Jeannette
1946. *This Magic Body.* New York: Viking Press.
Loren, Teri
1978. *The Dancer's Companion.* New York: Dial Press.
Lynham, Deryck
1950. *The Chevalier Noverre: Father of Modern Ballet.* London: Sylvan Press.
Matt, Pamela
1993. *A Kinesthetic Legacy: The Life and Works of Barbara Clark.* Tempe, AZ: CMT Press.

1996. The Nature of Ideokinesis and Its Value for Dancers. In *Dance Kinesiology* (Ed. Sally Fitt). New York: Schirmer Books, pp. 335–41.

Noverre, Jean-Georges

1930[1760]. Letters on Dancing and Ballets. (Trans. Beaumont). London: C. W. Beaumont [Originally published in 1760 as *Lettres sur la Danse et sur les Ballets*].

Radell, Sally, with Daniel Adame and Steven Cole

2004. The Impact of Mirrors on Body Image and Classroom Performance in Female College Ballet Dancers. *Journal of Dance Medicine and Science* 8(2): 47–52.

Raiport, Grigori

1988. *Red Gold: Peak Performance Techniques of the Russian and East German Olympic Victors.* Los Angeles: Tarcher.

Richardson, Alan

1967a. Mental Practice: A Review and Discussion I. *Research Quarterly of the American Association of Health and Physical Education* 38: 95–107.

1967b. Mental Practice: A Review and Discussion II. *Research Quarterly of the American Association of Health and Physical Education* 38: 263–93.

1969. *Mental Imagery.* London: Routledge and Kegan Paul.

Sparger, Celia

1982[1965]. *Anatomy and Ballet.* New York: Theatre Arts Books. [Fourth edition, printed in 1965, published by Adam and Charles Black, London].

Suinn, R. M.

1976. Visual Motor Behavior Rehearsal for Adaptive Behavior. In *Counselling Methods* (Ed. J. Krumboltz and J. Thorensen). New York: Holt, pp. 360–66.

Sweigard, Lulu

1938. Body Mechanics and Posture in Modern Life. In *Symposium on Posture,* printed as a contribution to the profession of Physical Education by Phi Delta Pi (National Professional Physical Education Fraternity for Women), March, pp. 18–27. Available in the archives of Springfield College, Springfield, MA.

1939. Bi-Lateral Asymmetry in the Alignment of the Skeletal Framework of the Human Body. PhD thesis, School of Education, New York University.

1946. Constructive Rest. *Self Magazine* 1(4): 21–22.

1961. The Dancer's Posture. *Impulse: The Annual of Contemporary Dance,* volume and series unknown. San Francisco, California, pp. 38–43.

1965. Better Dancing through Better Body Balance. *Journal of Health, Physical Education and Recreation* 36(5): 22–23, 56. May.

1974. *Human Movement Potential: Its Ideokinetic Facilitation.* New York: Dodd, Mead.

Todd, Mabel

1937. *The Thinking Body: A Study of the Balancing Forces of Dynamic Man.* Boston: Charles T. Branford.

1976[1953]. *The Hidden You: What You Are and What to Do about It.* New York: Dance Horizons [1953 edition published by Exposition Press, also in New York].

Urciuoli, Bonnie

1995. The Indexical Structure of Visibility. In *Human Action Signs in Cultural Con-*

text: The Visible and the Invisible in Movement and Dance (Ed. Brenda Farnell). Metuchen, NJ: Scarecrow Press, pp. 189–215.

Williams, Drid

1980. Taxonomies of the Body. *Journal for the Anthropological Study of Human Movement* [JASHM] 1(1): 1–19, and 1(2): 98–122.

1981. On *Human Movement Potential*: A Review Article. *JASHM* 1(4): 288–93.

1982. "Sweigard: A Memorial Tribute," in a program designed for the honoring of past New York University graduates who distinguished themselves in various fields.

2000. The Cultural Appropriation of Dances and Ceremonies. *Visual Anthropology* 13: 345–62.

2004. *Anthropology and the Dance: Ten Lectures*. Champaign: University of Illinois Press.

Author Index

Subject Index

Based in Minnesota, DRID WILLIAMS is the senior editor of the *Journal for the Anthropological Study of Human Movement* and the author of *Anthropology and the Dance: Ten Lectures*; *Anthropology and Human Movement: The Study of Dances*; and *Anthropology and Human Movement: Searching for Origins*. Before becoming an anthropologist, she was a dancer, teacher of dancing and choreographer for twenty years.

The University of Illinois Press
is a founding member of the
Association of American University Presses.

Composed in 10.5/13 Adobe Minion Pro
at the University of Illinois Press
Manufactured by Cushing-Malloy, Inc.

University of Illinois Press
1325 South Oak Street
Champaign, IL 61820-6903
www.press.uillinois.edu